T0318811

Nonlinear Kalman Filtering for Multi-Sensor Navigation of Unmanned Aerial Vehicles

Series Editor
Jean-Paul Bourrières

Nonlinear Kalman Filtering for Multi-Sensor Navigation of Unmanned Aerial Vehicles

Application to Guidance and Navigation of Unmanned Aerial Vehicles Flying in a Complex Environment

Jean-Philippe Condomines

First published 2018 in Great Britain and the United States by ISTE Press Ltd and Elsevier Ltd

ISTE Press Ltd
27-37 St George's Road
London SW19 4EU
UK

www.iste.co.uk

Elsevier Ltd
The Boulevard, Langford Lane
Kidlington, Oxford, OX5 1GB
UK

www.elsevier.com

British Library Cataloguing-in-Publication Data
A CIP record for this book is available from the British Library
Library of Congress Cataloging in Publication Data
A catalog record for this book is available from the Library of Congress
ISBN 978-1-78548-285-4

Printed and bound in the UK and US

Contents

Preface

The use of Unmanned Aerial Vehicles (UAVs) is exploding in the civil sector. With a market expected to exceed two billion euros by 2015 in France alone, 345 UAV operators had already registered a total of 585 aircraft with the Directorate General for Civil Aviation (DGAC) by late October 2012. Although UAVs are old news in the military sector, they are a brand new field for civil applications, such as pipeline monitoring, public protection or tools for processing and analyzing crops. New applications that use UAVs as experimental vectors are currently being researched. Among many other possible applications, UAVs seem especially promising in the fields of aerology and meteorology, where they can be used to study and measure local phenomena such as wind gradients and cloud formations. Interestingly, since 2006, mini-UAVs and micro-UAVs account for most new aircraft. Both belong to the category of sub-30-kg UAVs, which will be the primary focus of this book. These aircraft have the advantage of being relatively lightweight and easy to transport, unlike other types of UAV, which can weigh over 150 kg. Other than weight, UAVs can be classified by battery life, which determines their operating range. Accordingly, they are often categorized as Short Range (SR), Close Range (CR) or Medium Range (MR) aircraft. Although design configurations can vary wildly, the UAVs in any given category tend to share the following characteristics: (1) take-off weight, empty weight, nominal weight and size – these parameters create strong constraints on the maximum number and performance of the UAV's on-board sensors; (2) battery life and maximum range – these parameters determine the applications for which the UAV is most suitable; (3) flight parameters with various degrees of uncertainty and compatibility with scenarios such as flying indoors, in cluttered environments, against the wind, etc. More generally than the specific context of UAVs, a common approach to optimizing the performance of an avionic system is to establish specifications in terms of autonomy properties and closed-loop flight characteristics that satisfy the expected mission requirements of the aircraft.

Concretely, we can distinguish between the hardware components of an avionic system, which would typically consist of an embedded processor, sensors, an array of actuators and a ground/air communication module, and its software components, which would include the following:

– signal processing algorithms for a wide range of functions, such as denoising the sensor outputs or estimating and reconstructing the state of the aircraft or other flight parameters by merging all on-board measurement data (which often have low levels of redundancy), corroborated against the output of a predictive mathematical model describing the dynamic behavior of the aircraft;

– "low-level" control algorithms for closed-loop operation and guidance of the aircraft, allowing it to be programmed with instructions;

– "high-level" control algorithms for navigation, rerouting or decision support (AI) in the absence of a human operator, or in suboptimal scenarios (loss of comms, mechanical failure, etc.).

Thus, control algorithms clearly have an essential role to play, as do the algorithms that estimate the state or parameters of the aircraft, especially since cost and space constraints limit the capacities of the underlying sensor and actuator technologies. This is particularly relevant for micro-UAVs and mini-UAVs. Estimation algorithms allow us to merge imperfect information obtained from different sensors in real time in order to construct estimates, e.g. of the state of the UAV (orientation, velocity, position), by running control algorithms on the on-board processor. The closed-loop controls need to guarantee that the UAV remains stable regardless of the order in which instructions are received from the operator or automatic flight management system, while also ensuring that all instructions were received correctly. Estimation and control are therefore crucial aspects of every mission. One extremely important dimension of the mini-UAVs discussed in this book is their payload capacity. Mini-UAVs have relatively limited space. Combined with the budgetary constraints of mini-UAV development projects, this ultimately means that only so-called “low-cost” equipment is viable. Despite significant progress in miniaturization and a steady growth in on-board processing power (see Moore’s law), these mini-UAVs can therefore only realistically use limited-performance sensors to accomplish the ever-expanding panel of missions with which they are entrusted. For these new missions, mini-UAVs must be able to safely enter and share civil airspace; they must be able to pass flight certifications equivalent to those imposed on cargo flights operated by commercial airlines. In the context of safety, using estimation techniques to consolidate the UAV’s on-board knowledge of its own state becomes an essential component of the control framework, especially in suboptimal operating conditions (sensor failure, intermittent loss of signal, noise and perturbations from the environment, imperfect measurements, etc.). Attempting to tackle these challenges has naturally led researchers to explore relatively new problems, some of which are quite different from those encountered in civil and military aeronautics, whose avionic systems can differ drastically from those considered in this book. In our case, we need extremely sophisticated avionic systems that must perform a diverse range of on-board functions with minimal weight or space usage. In particular, our estimation algorithms must satisfy a number of very strong constraints, not only in terms of performance, but also execution time, memory space and convergence properties.

This book presents an original algorithmic solution to the problem of estimating the state of a mini-UAV in flight in a manner that is compatible with the inherent payload constraints of the system. Our approach is oriented toward model-based nonlinear estimation methods. Our first step is to define a dynamic model that describes the flight of the mini-UAV. This model should be sufficiently general as to work with multiples types of mini-UAV (fixed-wing, quadcopter, etc.). Next, two original estimation algorithms, called IUKF and π-IUKF, are developed on the basis of this model, then tested, first against simulations, then against real data for the π-IUKF algorithm. These two methods apply the general framework of invariant observers to the nonlinear estimation of the state of a dynamic system using an Unscented Kalman Filter (UKF) method, from the more general class of Sigma Point (SP) nonlinear filtering algorithms. In future, the solutions outlined here will be integrated into the avionics of the mini-UAVs studied at the ENAC laboratory; the source code is already available as part of the autopilot of the Paparazzi project.

Organization of this book

This book is divided into five chapters:

– Chapter 1 presents the background of aerial robotics at the time of writing, giving an overview of the various technological advancements that have allowed it to experience such extensive growth in the civil sector. The growth of aerial robotics has inspired a variety of so-called "open source" development projects, which aim to provide a comprehensive autopilot system for mini-UAVs. In the conclusion of this chapter, we explain the motivation behind the research presented throughout the rest of the book and discuss the challenges associated with designing a mini-UAV state estimator that is compatible with the inherent capacity constraints of the open-source Paparazzi framework.

– Chapter 2 presents the latest advancements in estimation techniques. This literature review is used as a reference throughout the rest of the book, and focuses in particular on two specific techniques: Kalman filtering and invariant observers. Any readers who are unfamiliar with differential geometry can additionally refer to Appendix A. It is based on a combination of these two techniques that we were able to develop and validate our own two nonlinear estimation algorithms.

– In Chapter 3, we outline the various kinematic models commonly used to manage the navigation of dynamic systems. The estimator filters developed in this book were built from these more general models, which allowed us to account for perturbations to the system in the form of random errors. We conducted a detailed observability study to determine whether the state variables can be reconstructed from the known system inputs and measurements. We shall see that accurate dynamic models of inertial navigation can have several degrees of freedom, and so a certain number of model assumptions are required to guarantee that the estimation problem remains observable.

– Based on the models established in Chapter 3, Chapter 4 documents the development of a set of original methodological principles that allowed us to construct two nonlinear estimation algorithms, IUKF and π-IUKF, which differ in terms of their formulation. These algorithms, founded on the theories of invariant observers and so-called "unscented" Kalman filters, offer an extremely valuable algorithmic solution to the challenges of inertial navigation. By comprehensively summarizing the initial results which characterize the IUKF algorithms, we demonstrate their well-foundedness and highlight their advantages, both theoretical and practical.

– Chapter 5 presents all of the results obtained so far regarding the π-IUKF algorithm. The first part of this chapter compares the performance of the SR-UKF (standard) and π-IUKF algorithms. The analysis is based on simulated noisy data generated from the general models introduced in Chapter 3 after accounting for imperfections in each type of sensor. Continuing with the case of a complete navigation model, the second part of this chapter presents several experimental results obtained by estimating the state of the mini-UAV from real data using the three primary algorithms considered: SR-UKF, π-IUKF and the classical invariant observer approach. These results validate the approach, allowing a specific correction to be derived for the prediction obtained from any given representation of a nonlinear state used in the estimation, in such a way that the dynamics of the constructed observer satisfy the symmetry properties of the system.

Finally, the book is brought to a close by a collection of appendices, which are referenced wherever relevant throughout the rest of the book.

Jean-Philippe CONDOMINES
June 2018

1

Introduction to Aerial Robotics

1.1. Aerial robotics

Aerial robots, also known as Remotely Piloted Aerial Systems (RPAS) or Unmanned Aerial Vehicles (UAVs), are unmanned aircraft that can complete their mission with some degree of autonomy. Their primary purpose is to execute a task more safely or effectively than a remotely-controlled aircraft. The possibilities offered by autonomous systems such as UAVs in the civil sector have been thoroughly explored over the last few years. Various research projects, including some financed by the European Commission, have studied the potential civil applications of UAVs. Similar aircraft had previously been frequently used for specific military purposes in various interventions, including in Iraq and Afghanistan, where they played a key role as active links of information, and in decision and action networks. Today, the operational benefits provided by UAVs with access to sufficient decisional resources have revolutionized the intervention scenarios of missions conducted in hostile zones with significant risk to human life. Surveillance missions have undergone a similar metamorphosis and are completely unrecognizable compared to just a few decades ago. The system now assumes responsibility for piloting and guidance tasks, as well as lookout tasks for which human vigilance has proven fallible. UAVs can now relieve their human operators, allowing the latter to dedicate more time to managing the mission at a higher level. UAVs are also currently being studied as experimental vectors for various applications.

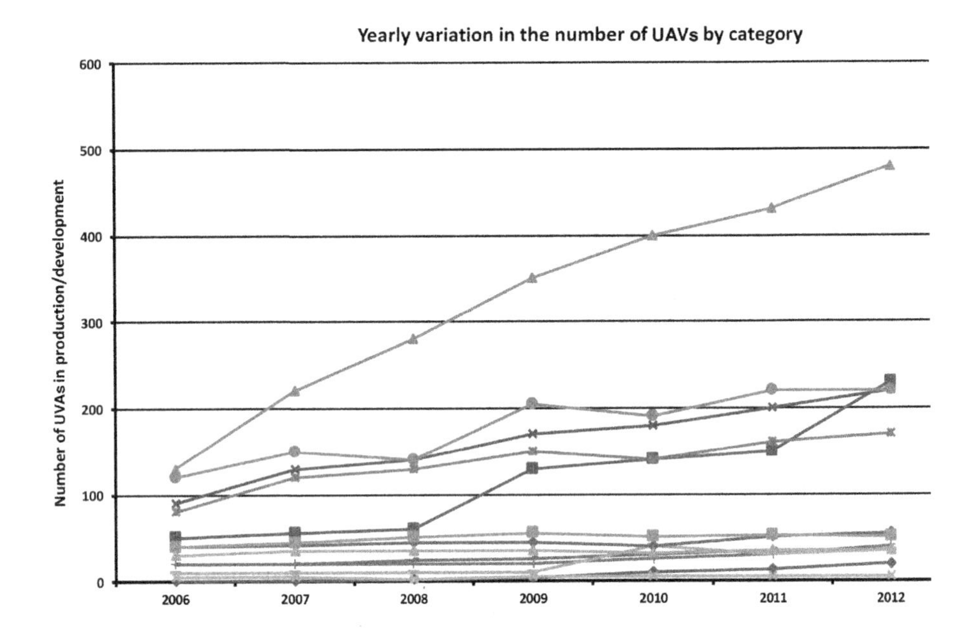

Figure 1.1. *Number of UAVs in production and development from 2006 to 2012. For a color version of this figure, see www.iste.co.uk/condomines/kalman.zip*

1.1.1. *The rise of UAVs in the civil sector*

Since 2006, the growth in UAVs has been strongest in the categories of mini- and micro-UAVs, both of which are types of sub-30 kg UAV (see Figure 1.1, *www.sesarju.eu*), the primary focus of this book. The UAVs in this category have the advantage of being relatively lightweight and easy to transport, unlike other types of UAV, which can weigh over 150 kg. Aerial robots exist in an extremely wide variety of forms. Other than weight, they can be classified by endurance, which determines their range. For example, we can distinguish between High-Altitude Long Endurance (HALE) and Medium-Altitude Long Endurance (MALE) UAVs, as well as so-called medium- and short-range UAVs, and mini-UAVs. The classification that we will use in this book is based on the one proposed by the US Air Force. RPASs can also be characterized by their applications, according to their original military purposes. For example, we can distinguish between strategic UAVs, tactical UAVs and combat UAVs (Unmanned Combat Air Vehicles [UCAVs]). These UAVs can have fixed or rotary wing configurations, or even hybrid flight systems. Strategic UAVs are usually HALEs and are typically used for reconnaissance missions. Some have a battery life of several days and can fly at altitudes of over 20,000 m. The two best-known models of HALE are the Global Hawk, manufactured by Northrop Grumman and the Sentinel, developed by Lockheed Martin (see Figure 1.2).

(a) RQ-4B Global Hawk

(b) RQ-170 Sentinel

Figure 1.2. *Examples of HALE UAVs*

Tactical UAVs can take various forms, ranging from MALEs to mini-UAVs, or even micro-UAVs. With a battery life of around 30 h, MALE UAVs can be used for an extremely broad range of military or civil missions, usually flown at altitudes of between 5,000 and 15,000 m. Notable examples of such UAVs include the Reaper and the Gray Eagle, both developed by General Atomics (see Figure 1.3).

(a) MQ-9 Reaper

(b) MQ-1C Gray Eagle

Figure 1.3. *Examples of MALE UAVs*

These two categories of UAVs can be clearly distinguished from UCAVs, whose technical characteristics are very different (only a dozen hours of battery life and flown at less than 5,000 m). Thus, combat UAVs have significantly lower endurance, but have take-off weights that are comparable to classic combat aircraft (around 10 tons compared to 1.2 tons for the

Predator A). Medium-range (MR) and short-range (SR) tactical UAVs are designed specifically for surveillance missions and are heavily exploited in civil environments, for example as communication relays. The development of this type of UAV is currently in rapid expansion around the globe, including in France. One notable example is the Sperwer model by Sagem.

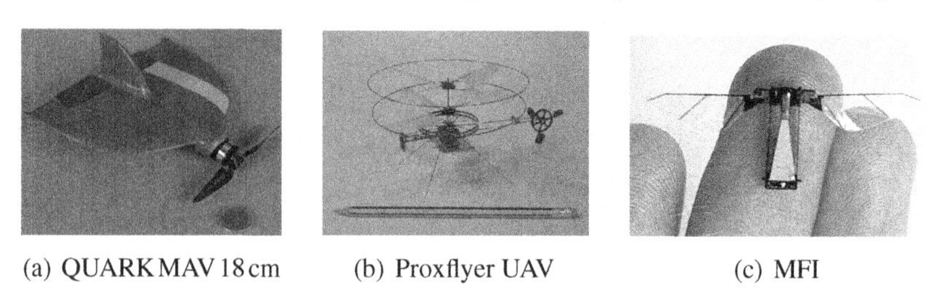

(a) QUARKMAV 18 cm (b) Proxflyer UAV (c) MFI

Figure 1.4. *Examples of micro-UAVs. For a color version of this figure, see www.iste.co.uk/condomines/kalman.zip*

Finally, as their name suggests, mini-UAVs are very small UAVs. They are usually around a meter in size and are characterized by a battery life of a few hours. This category tends to be the most useful for developing easy-to-implement and inexpensive experimental platforms. Among other things, mini-UAVs are used to illustrate and demonstrate research in the fields of robotics, automation and signal processing. Mini-UAVs played a key role in our development and implementation of a series of crucial estimation algorithms for UAVs, in part because the capacities of sensor and actuator technologies are limited by cost and space constraints. Figure 1.4 shows some of the various designs that currently exist for miniaturized aerial vehicles: the Quark UAV developed by the automation group of the MAIAA team at the ENAC laboratory in France, one example of a UAV from the Proxflyer range and finally an example of a Micromechanical Flying Insect (MFI) UAV.

1.1.2. *Mini-UAV designs*

Rotary wings: UAVs with multiple rotors or rotary wings are the most widely recognized aeromechanical configuration and the most common design for autonomous aerial vehicles. They typically have four rotors, but some models can have six or even eight. This design is especially good for stationary and low-speed flying, but is very poorly optimized for extended flights (maximum of around 30 min, depending on the weight of the aircraft), and high-speed lateral flying is not recommended.

Fixed wings: Long favored by radio-controlled model enthusiasts, fixed-wing UAVs are mostly used for observation missions. By taking advantage of cells that are energetically more efficient than those used by rotary-wing UAVs, they can achieve flight times of over an hour. Unfortunately, the endurance of these cells diminishes with size. Various studies on high-endurance mini-UAV design have been conducted to optimize each component of the vehicle (motor system, propellers, wings, fuselage, etc.) in order to combine the benefits of both mini-UAVs and UAVs with larger wingspans.

Transitioning vehicles: Although still a relatively niche category, convertible UAVs offer an alternative solution to the biggest issues with fixed-wing and rotary-wing designs by allowing stationary and very low-speed flight to be combined with lateral flight. Several studies have been performed on vehicles that behave like a helicopter at low speeds and an airplane at high speeds. In practice, these aircraft typically involve mechanically complex designs, with results that are extremely structurally fragile. Another problem is that the transition between helicopter phase and airplane phase is critical, yet its aerodynamics remain poorly understood.

1.2. The Paparazzi project

The historical development of aerial robotics was shaped by contributions from a very large number of groups, from both academia (research laboratories, universities, institutions, etc.) and the private sector. As a result, there has been extensive focus on the following areas: the design of innovative aircraft (especially mini-UAVs), the commercialization of sensors with various cost and performance properties, and the development of firmware to greatly expand the abilities of aerial robots, allowing them to complete a wide range of missions. To cite just a few examples, the PIXHAWK project by ETH Zurich studied vision for autonomous aerial robots; the STARMAC (Stanford/Berkeley Testbed of Autonomous Rotorcraft for Multi-Agent Control) platform by Stanford/Berkeley University perfected new control designs for multiagent systems; the German company Ascending Technologies commercialized several fully equipped mini-UAVs with preprogrammed autopilot chips and sold together with development tools; and the French company Parrot commercialized the ARDRONE-2, one of the first consumer UAVs. The impressive growth of the field of aerial robotics over the past 10–15 years has also inspired a number of so-called "open source" development initiatives, passion projects by enthusiasts from various backgrounds (researchers, students, radio-controlled model enthusiasts, electronics engineers, software developers and much more). The work presented in this book follows a similar trend, intended as a contribution to an open-source system called Paparazzi[1] [HAT 14] whose objective is to offer a comprehensive autopilot system for mini-UAVs. Open-source projects rely on contributions from a community of developers who share all of their progress and methodological, technological and programming works.

Each community contribution is evaluated by feedback from hundreds of other users throughout the world. A vast number of other projects are organized in this way, the most prominent of which is arguably the Linux operating system. In the field of robotics, around 2,000 projects [LIM 12] have adopted open-source structures. This includes alternatives to Paparazzi, such as OpenPilot, Arducopter, Pixhawk by ETH Zurich, MikroKopter by HiSystems GmbH for software components and KKMultiCopter, Multiwii and Aeroquad for other components. Most mini-UAV development projects in both academia and the private sector rely on relatively general-purpose estimation algorithms known as Attitude and Heading Reference Systems (AHRS) or Inertial Navigation Systems (INS). These algorithms simply estimate the attitude and in some cases also the position and speed of the aircraft. As we will see in Chapter 2, modern systems can choose from a large selection of methods for estimating these key variables to determine the state of the mini-UAV. Each of the open-source projects cited above proposes multiple relatively standard solutions to the estimation problem. For each open-source project, Figure 1.6 lists the filtering algorithms used by the autopilot of the aircraft and the navigation modes available in both of the mono- and multicases, as well as the automatic control functionality supported by the project. Each control mode regulates the mini-UAV's ability to maintain a certain position, attitude or heading using its sensors. In order for sensor-based control to be possible, a characterization of the imperfections of each on-board measurement system is required, so that an efficient filtering

1 www.paparazzi-uav.org.

algorithm can be developed. We will discuss the possible characterizations section 1.3. Additionally, the choice of estimation method sometimes needs to be compatible with certain properties of the navigation model. Two different estimation algorithms are used by the projects listed in Figure 1.6, each with its own advantages and drawbacks: (1) the Extended Kalman Filter (EKF) algorithm, presented in Chapter 2, has the disadvantage of using a locally linear model around the estimated trajectory, which can lead to non-convergence if the model does not represent the dynamics of the UAV sufficiently well at the relevant point in its flight. If the linearization can be established analytically, this algorithm requires relatively low computation time, but it can take much longer if numerical calculations are required. However, EKF has the benefit of using a relatively comprehensive model of the flight kinematics and dynamics, and is effective at eliminating noise; (2) linear/nonlinear complementary filters (LCF/NCF) [MAH 08] can only be applied to part of the navigation model (AHRS), which makes the estimation process sensitive to loss of GPS signal. Combining this approach with other sensors has proven impossible in practice. The benefit of NCF filters is that they provide a highly accurate estimate of the attitude of the aircraft.

The architecture of the Paparazzi system is characterized by a data acquisition loop, a system state estimation step and a command block that the UAV uses to follow the flight plan specified by the user (see Figure 1.5). Within this overarching structure, which was established in the earliest stages of Paparazzi, the estimation step is positioned between the data acquisition loop (dark-green block) and the command calculation (stabilization block). Estimation is performed according to a classical approach. The estimation algorithm aggregates information from all available sensors in order to improve the performance of the command and navigation loops in real time. In Paparazzi, these loops consist of three independent filters, one for each of the following three subsystems, serving to estimate the navigation variables or the attitude of the UAV (INS/AHRS blocks):

– *an AHRS (NCF)* merges data from the inertial sensors (accelerometers and gyroscopes) to estimate the attitude of the aircraft with respect to its three axes: roll, yaw, pitch;

– *a so-called vertical Kalman filter* combines the scalar measurements from the barometer or the vertical component of the GPS and the acceleration to estimate the altitude of the aerial vehicle;

– *the third subsystem* simply takes the reading of the GPS signal as an input, allowing the UAV to calculate its position in the horizontal plane.

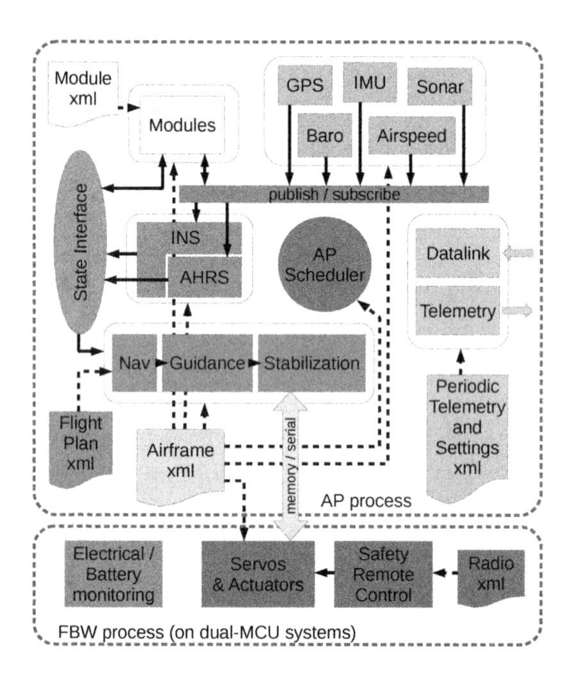

Figure 1.5. *Paparazzi software architecture. For a color version of this figure, see www.iste.co.uk/condomines/kalman.zip*

1.3. Measurement techniques

Every sensor used in mobile robotics, whether for terrestrial, aquatic or aerial purposes, can be classified into the following two categories [ALD 01]:

– *Proprioceptive sensors* measure the movement of the vehicle through space based on locally perceived information (e.g. acceleration measurements). This produces results that are easy to manipulate and do not depend on the environmental conditions, which play a key role in autonomous UAV navigation. However, as we will see below, with so-called "low-cost" technology, the accuracy of this information deteriorates gradually over time, rendering it unreliable and possibly even unusable as a unique reference in the long term. Any such limited-performance sensors will therefore need to be complemented by a source of additional information.

	Openpilot www.openpilot.org	Pixhawk www.pixhawk.org	Aeroquad www.aeroquad.com	Arducopter www.diydrones.com	Paparazzi wiki.paparazziuav.org	Mikrokopter www.mikrocopter.de	Multiwii www.multiwii.com
Filters & models							
Estimation	EKF	EKF	NCF/KF	NCF/KF	NCF/KF	LCF/KF	LCF/KF
Complete INS model	✓	✓	—	—	—	—	—
Special features							
Multi-cell platform	✓	—	—	✓	✓	✓	—
GPS-based navigation with checkpoints	Δ	Δ	—	✓	✓	Δ	Δ
Vision-based indoors flight	—	✓	—	—	—	—	—
Flight control (type of sensor)							
Heading control (magnetometer)	✓	✓	✓	✓	✓	✓	✓
Position control (GPS)	Δ	Δ	—	✓	✓	Δ	Δ
Altitude control (barometer)	✓	Δ	✓	✓	✓	✓	✓
Altitude control (GPS)	Δ	Δ	—	✓	✓	Δ	Δ
Altitude control (sonar)	—	—	✓	✓	—	—	—
Altitude control (optical flux camera)	✓	✓	—	—	—	—	—
Open software license	GPL	GPL	GPL	LGPL	GPL	—	GPL

Δ: partially used (e.g. other components required for navigation); ✓: used; – not used.

Figure 1.6. *Comparison of open-source projects in 2010. GPL – General Public License, LGPL – Lesser General Public License. For a color version of this figure, see www.iste.co.uk/condomines/kalman.zip*

– *Exteroceptive sensors* collect information about the environment of the vehicle (pressure, magnetic field, etc.). This information consists of observations about an absolute reference system attached to the environment (e.g. satellite position). These sensors produce invaluable measurements that can be used to periodically recalibrate the localization obtained by using proprioceptive sensors to measure the vehicle's motion.

1.3.1. *Proprioceptive sensors*

1.3.1.1. *Gyroscopes*

By definition, a gyroscope (or gyrometer) is an instrument that measures the angular velocity of a moving object along one or several axes of a coordinate system with respect to an inertial frame of reference. There are two large families of gyroscope: optical gyroscopes and mechanical gyroscopes. The former are based on the Sagnac effect [LAW 98], and the latter are based on the Coriolis effect [TIT 04]. The Sagnac effect is a physical phenomenon exploited by laser gyroscopes and optical fiber gyroscopes. These types of sensors are usually extremely effective, but they tend to be bulky. The Coriolis effect, also known as the Coriolis force, is used by piezoelectric oscillation gyroscopes (see Figure 1.7(c)). The principles of this effect have been understood since Foucault's famous pendulum experiment in 1851. These gyroscopes work by measuring the coupling between two modes of vibration (excitation and detection) induced by the Coriolis effect. The sensors contain a large number of vibrating elements that are kept in continuous motion by sinusoidal excitations (excitation mode). When the system is at rest, the resultants of the signals that propagate through the vibrating elements are identical and have constant amplitude. However, when the system is rotating, perturbations caused by the Coriolis force create amplitude differences that are proportional to the speed of rotation (detection mode).

The most important parameters that influence the accuracy of a gyroscope are the bias, the scale factor (or sensitivity) and the measurement resolution of the sensor. The bias is defined as the modification or drift in the output signal of the sensor (expressed in volts) when there is no input signal. The scale factor is defined as the ratio of the sensor output to the sensor input (which is expressed in degree/sec); the scale factor has units of volt/degree/sec. Any change in the bias over time or due to temperature variation directly induces an absolute measurement error (expressed in units of degree/sec), whereas a change in the scale factor creates a relative error (expressed in units of parts per million [ppm]). To gain an understanding of what these units of ppm represent, consider that an error of 300 ppm (0.03%) in the scale factor leads to an angular error of 0.1 after one full rotation of the gyroscope. Finally, the sensor resolution (expressed in degree/sec) is defined as the smallest measurable input signal. The resolution is the product of the spectral density of noise (expressed in degree/sec/$\sqrt{\text{Hz}}$) and the square root of the bandwidth of the sensor (expressed in Hz).

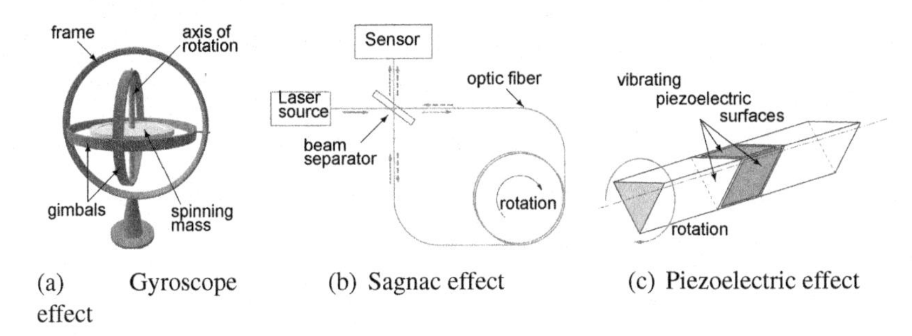

(a) Gyroscope effect (b) Sagnac effect (c) Piezoelectric effect

Figure 1.7. *Examples of gyroscopes*

When choosing which gyroscope technology to integrate into an autonomous aircraft, each of the properties described above must be carefully weighed to ensure that the UAV can fulfill its mission requirements. Figure 1.8 gives an overview of typical choices of gyroscope accuracy parameters (bias and scale factor) for specific types of missions based on current technology [TIT 04].

Newly emerging technologies, such as those based on microelectromechanical systems (MEMS), are clearly superior to inertial sensors based on interferometry, at least in terms of weight and size. Originally used for military applications, MEMS technologies are now accessible to the general public because of their widespread adoption by mobile telephone providers, which led to a decrease in production cost. This technology has significantly reduced the size of gyroscopes. Over the next few years, the military market will likely incentivize the development of this technology for applications such as intelligent ammunition or guidance systems for tactical missiles with short flight times. Used in a wide variety of applications, MEMS and IFOG (interferometry) technologies can be expected to eventually replace many of the current inertial systems based on ring laser gyroscopes (RLGs) and mechanical masses. Note that some fields, like aquatic robotics, prefer RLG technology due to the presence of strong constraints on the stability of the scale factor; in this specific regard RLGs strongly outperform IFOGs.

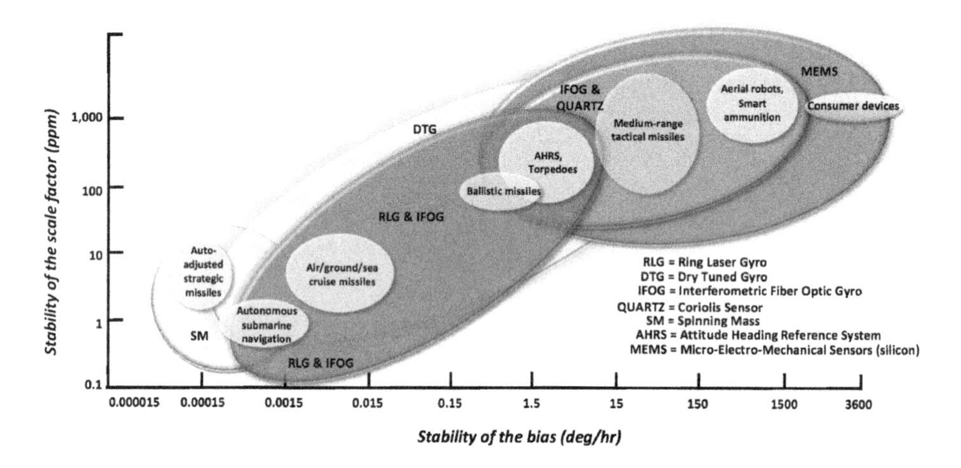

Figure 1.8. *Technological solutions for specific mission requirements. For a color version of this figure, see www.iste.co.uk/condomines/kalman.zip*

1.3.1.2. *Accelerometers*

As its name suggests, an accelerometer measures the linear acceleration of any moving or stationary object to which it is attached. Physically, accelerometers contain a test mass (also known as a seismic mass) with accurately known weight and shape, suspended by some mechanism (spring, flexible strip, membrane) from a rigid frame. At rest, or when the object is moving at constant linear speed, the mass is in equilibrium, as shown in Figure 1.9(a). By contrast, whenever the sensor experiences sudden non-zero acceleration, the test mass begins to move, acted upon both by this sudden variation and the restoring force from the spring. By measuring its displacement, we can calculate the acceleration, which is directly proportional to the measured extension: $m\ddot{x} + kx = 0$. In Figure 1.9(b), the object is experiencing an acceleration directed toward the right. The measuring space is bordered by two blocks that limit the range of the seismic mass to avoid damage to the spring. This limit defines the maximum acceleration that the sensor can withstand before sustaining permanent damage. Various techniques have been proposed in the literature to measure the displacement of the test body and transform the action of the acceleration on the sensor into an electronic signal. This type of sensor is characterized by multiple parameters, including the operating range, expressed in ms^{-2}, the bandwidth (expressed in Hz), the resolution (expressed in g) and the scale factor (sensitivity), which provides information about the detection quality for small variations in the input around the measured value (expressed in V/g or in relative units [ppm]), and the maximum tolerated shock (expressed in g). Most SI documentation recommends that acceleration should be measured in units of $g \approx 9.81\,\mathrm{ms}^{-2}$, so that one unit is equal to the acceleration due to gravity. As before, the choice of accelerometer technology is made according to the mission requirements of the UAV. Figure 1.10 gives an overview of the most important parameters (the bias and the scale factor) that characterize the accuracy of an accelerometer in the context of the various types of mission that UAVs can accomplish with modern technology. As shown by the figure, civil and military applications are once again dominated by MEMS technology, since it is usually low cost, and has not yet been superseded (quartz resonators are more expensive and provide similar quality).

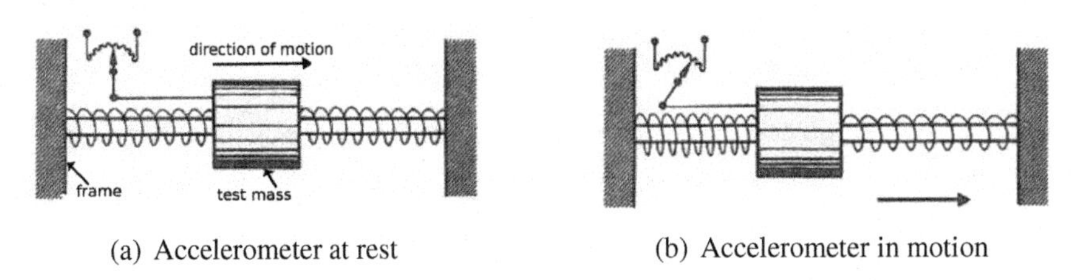

(a) Accelerometer at rest (b) Accelerometer in motion

Figure 1.9. *Principle of the accelerometer*

These kinds of questions about the load-bearing capacity of mini-UAVs are an extremely important dimension that have guided much of the work presented throughout this book. UAVs have a relatively limited capacity. Together with the development budget constraints associated with any mini-UAV project, this limits the choice of materials and components to so-called "low-cost" options. Figure 1.11 gives a brief overview of the price of each of the proprioceptive sensors presented above as a function of the underlying technology and the required performance [TIT 04]. Despite significant progress in miniaturization and the constant increase in the processing power of embedded systems (Moore's law states that κ doubles every 18 months), the mini-UAVs that we shall consider here can only feature sensors with limited performance, given that this category of autonomous aircraft is being called upon ever more frequently to accomplish a constantly expanding range of missions.

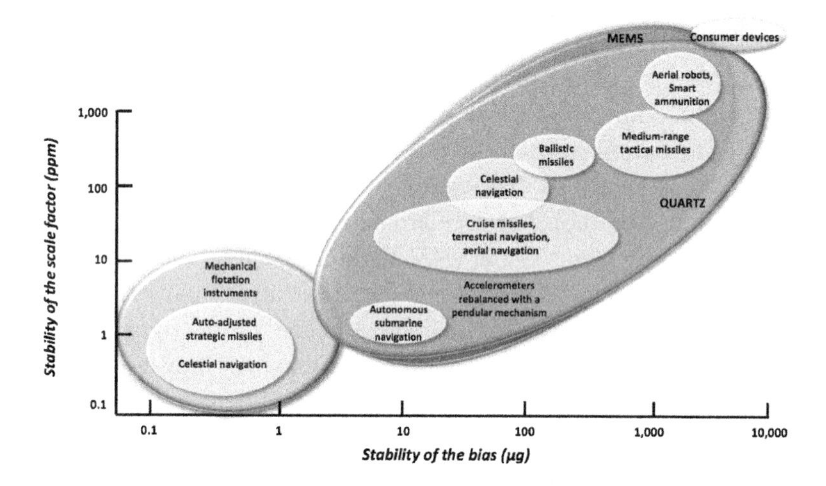

Figure 1.10. *Applications of accelerometer technologies. For a color version of this figure, see www.iste.co.uk/condomines/kalman.zip*

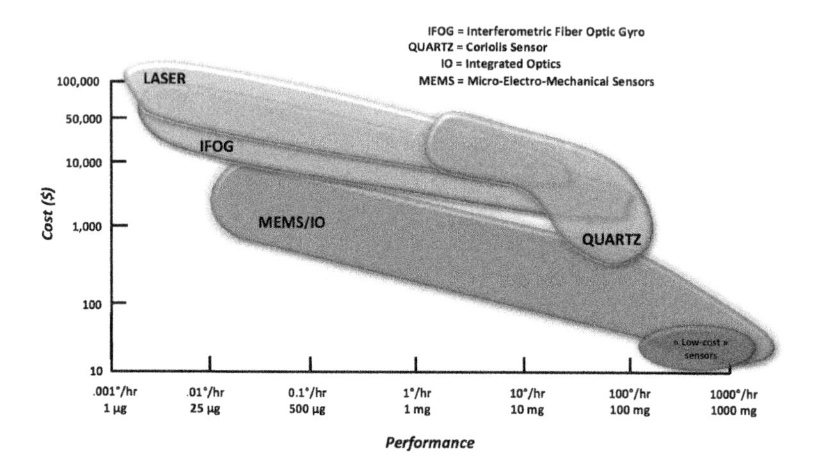

Figure 1.11. *Price of proprioceptive sensors by technology. For a color version of this figure, see www.iste.co.uk/condomines/kalman.zip*

1.3.2. *Exteroceptive sensors*

1.3.2.1. *Satellite positioning systems*

Satellite positioning systems can be used to determine the absolute position of an object on the Earth from a collection of signals broadcast by a constellation of satellites dedicated specifically to this purpose. By specifying the latitude, longitude and altitude, positions can be easily represented in space by a geodesic coordinate system ($WGS84$) up to an accuracy of around a dozen meters for most standard systems. These systems can therefore be used to locate transportation vehicles such as airplanes, ships, missiles or even low-orbit satellites.

Along with the Russian system GLONASS (GLObal NAvigation Satellite System), the best-known and most widely used satellite positioning system is arguably GPS (Global Positioning System). Developed and operated by the United States, GPS has been made available to the general public with deliberately reduced accuracy. To determine the absolute position of a receiver positioned somewhere on Earth, each satellite broadcasts one or more pseudo-random codes, always on two frequencies that are in phase relative to one another. These codes are precisely dated using an atomic clock on-board the satellite. Based on its own internal clock, the receiver can calculate the travel time of these signals between the broadcasting satellite and the receiver. Each distance measurement defines a sphere centered around the broadcasting satellite on which the receiver must be located. By combining the locations determined from three separate satellites, the absolute position of the receiver can be accurately deduced by trilateration. This principle is relatively simple, but cannot be implemented as such in practice. In order for the reasoning described above to work, the synchronization between the clocks of the three satellites and the receiver needs to be extremely precise; an offset of one thousandth of a second would lead to an error of over 300 km. To solve the clock synchronization problem, a fourth satellite is used, which fixes the broadcast time and thus eliminates any uncertainty. As well as clock synchronization errors, differences in the propagation speed of the signals through each layer of the atmosphere, reverberation and deliberate limitations imposed by the US military need to be taken into account. One way of mitigating these errors is to use a relative positioning system called Differential Global Positioning System (DGPS). This system, which is used by European Geostationary Navigation Overlay

Service (EGNOS), is based on a local positioning error calculated by a collection of fixed stations with perfectly known positions. This information is transmitted to the receiver by radio or by satellite. After 2020, it will be possible to improve the accuracy by using the GALILEO system, which is intended to allow the European Union to become strategically independent from the United States and Russia. In aerial vehicles, civil application of GPS can achieve an accuracy of around a dozen meters in open environments. However, urban locations tend to be cluttered, which introduces more randomness and means that a merging algorithm is required to improve the location-finding process. Another alternative is to use more powerful equipment, which is significantly bulkier and more expensive.

1.3.2.2. *Barometric altimeters*

A barometric altimeter is an instrument that measures the atmospheric pressure difference relative to some reference, usually the atmospheric pressure at sea level. In practice, these altimeters are typically used to find the altitude of an objective relative to a reference altitude based on the fact that the atmospheric pressure decreases as the altitude increases.

(a) GPS satellite

(b) GLONASS satellite

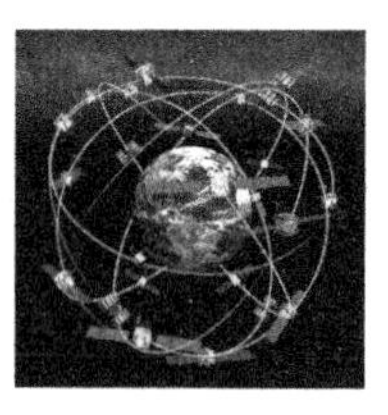

(c) Constellation of satellites

Figure 1.12. *Satellite-based positioning. For a color version of this figure, see www.iste.co.uk/condomines/kalman.zip*

Aneroid barometers are the most widely used technology for mini-UAV applications. The key component of these barometers is a steel or beryllium capsule containing a vacuum. Under atmospheric pressure, this capsule experiences an elastic deformation from which the pressure can be measured. However, precautions are required when using this measurement in calculations, since the capsule will also be deformed by strong accelerations and highly dynamic maneuvers. Readings from aneroid barometers are crucial for small UAVs, which require precise knowledge of their altitude at all times. However, these readings are also affected by the weather. Changes in the ambient pressure (anticyclone or depression) can lead to significant uncertainty in the measurements. Therefore, the confidence level given to this type of sensor varies relative to other sources of information about the UAV's altitude (such as GPS).

1.3.2.3. *Magnetometers*

A magnetometer or compass is a navigation instrument that can identify a specific reference direction (usually the north) in the horizontal plane, allowing horizontal angles to be measured with respect to this direction. Magnetometers can work in several ways. The simplest, the magnetic compass, also simply known as a magnetometer, tracks the orientation of a magnetic needle within the Earth's magnetic field, in the same way as a traditional compass. This is a natural solution that does not require a source of energy, which means that magnetic compasses are highly resistant to failure. However, magnetic north does not align with geographic north – the difference is called the magnetic declination. To adjust for this phenomenon, the compass must be constantly recalibrated with an offset to yield an accurate absolute measurement of the direction of geographic north. The biggest problem with these simple sensors is that they are highly sensitive to nearby magnetic objects and parasitic magnetic fields. This can corrected via calibration in the immediate environment of the UAV by statically quantifying the observed measurement offsets induced by magnetic bodies, with varying results. Electronic compasses, on the other hand, can find magnetic north from the electronic properties displayed by certain materials when a magnetic field is applied. These compasses can be based on various phenomena, such as the Hall effect, magnetoinduction or magnetoresistance. This type of sensor consists of a rigid frame that holds a flexible beam positioned such that the alternating current running through it passes above a compensation coil. When a magnetic field $\overrightarrow{B}$ is applied to the sensor, a magnetic force $\overrightarrow{F}$, known as the Laplace force, acts on the beam, inducing a mechanical deformation whose amplitude can be measured by a piezoresistive gauge bridge. By modulating the current through the compensation coil, any parasitic fields that might skew the results of the gauge can be compensated for. This type of technology can be used to construct magnetometers for electronic systems embedded on aerial mini-UAVs.

1.3.2.4. *Cameras and telemeters*

Cameras and telemeters are sensors that measure the distance or the orientation of an object using radio, acoustic or optical phenomena. In general, these sensors yield information about an object within the context of its environment. Telemeters work by transmitting a signal toward the object that needs to be located, then measuring the reflected signals. These instruments are also commonly used for three-dimensional cartography and detecting objects against a certain background. There are several types of telemetric sensors, based on different kind of waves. One category uses sound waves or SONAR (SOund Navigation and Ranging). This sensor technology, which operates at ultrasonic frequencies, is the most widely used for mobile and aerial robotics, due to its simplicity, compactness and low cost. However, these instruments tend to encounter major issues when measuring distances of more than a few centimeters. Beyond this range, objects can no longer be reliably detected, and false positives are frequent, due to the way that sound waves propagate. In practice, this technology is therefore only useful for measuring the distance between the UAV and the ground during the take-off and landing phases. Another category of sensors instead uses optical waves, called LIDAR (LIght Detection And Ranging) [TRU 98, MAN 12]. LIDAR works in essentially the same way as SONAR, but uses signals from a different spectral frequency domain. Cost is a significant obstacle for LIDAR technology, and its bulkiness also tends to be unsuitable for aerial robotics.

With regard to cameras, we now have several ways of extracting visual information from their data streams [MA 04, RUF 09]. One simple biomimetic approach inspired by bees extracts optical flux information from the image, i.e. the apparent displacement of the object relative to its background caused by the motion of the camera relative to its environment [RUF 04]. However, this technique, widely used for take-off and flight stabilization, can be undermined by the environmental conditions of the UAV (for example the wind might create undesirable relative motion in the image). Other approaches, such as cartography [MA 04, TRU 98, CYG 09], use a pair of cameras in the same way that humans use their eyes to estimate the position, orientation and speed of objects in space.

1.3.3. *Inertial systems*

Inertial systems can be divided into three categories: Inertial Measurement Units (IMU), Attitude and Heading Reference Systems (AHRS) and Inertial Navigation Systems (INS) (see Figure 1.13). By definition, an IMU is composed of three accelerometers and three gyrometers that measure the specific acceleration (excluding gravity) and the rotational velocity of a vehicle. By integrating these six measurements, the IMU is theoretically capable of calculating the linear velocity, the position and the orientation of the vehicle relative to its initial state [DEL 08] [GRE 01]. Most applications in mobile robotics therefore feature an IMU complemented by a satellite positioning system.

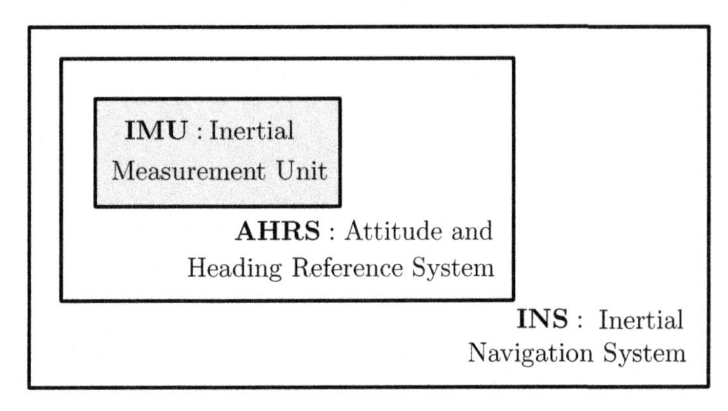

Figure 1.13. *Categories of inertial systems*

An AHRS is an IMU extended by real-time computations to determine the orientation of the object. Similarly, an INS is an AHRS that additionally computes the evolution of the velocity and position vectors of the vehicle from the components of the acceleration vector [FAR 98, FAR 08]. The primary difference between the latter two systems lies in the processing performed with the data extracted from the sensors. The sensors used by an INS and an AHRS are identical, but the algorithm of an AHRS does not compute the position. In practice, the theoretical capacities of IMUs are limited by imperfections in the acceleration and rotation measurements by the sensors (e.g. noise, bias, scale factors). Over time, these inaccuracies, which we discussed in section 1.3, lead to drift in the estimates of the position, the velocity and the orientation of the vehicle. There are however ways to limit the inaccuracy in these measurements. One possible approach is to require the error of any accelerometers and gyrometers to be less than microgram or a few hundredths of a degree per hour. Additionally, the measurements can be repeated at a rate of several hundred Hz to improve the accuracy. The cost of a high-performance INS can range up to hundreds of thousands of euros (see Figure 1.11). An alternative solution is to implement hybrid algorithms that merge information from proprioceptive sensors with information obtained separately from exteroceptive sensors. Merging algorithms can be used to calculate the position, velocity and orientation of a vehicle using measurements from a satellite positioning system, as well as the bias in the accelerometers and gyrometers. Thus, this approach is potentially capable of constructing a representative estimate of the true state of the vehicle at all times.

1.4. Motivation

Although the architecture implemented by the Paparazzi system is one of the most modular and development-friendly environments available for embedded algorithms (command, estimation, etc.), it currently does not offer any broad-scope data-merging features for processing the full set of on-board measurements. As noted in section 1.2, only a few data-merging operations, often limited to a single axis, are available. Thus, given the many-faceted challenges faced by aerial robotics (agile flight, urban zones, cluttered zones, possible loss of GPS signal, etc.), the functionality of the Paparazzi autopilot is insufficient, especially when reconstructing the state of the mini-UAV, which is required in order to develop advanced control laws. The root cause can ultimately be traced back to cost if we consider the underlying technologies used by this type of UAV. These observations motivated the work performed and presented in this book. It seems clear that the increasing flight capacities of the mini-UAVs used for research by ENAC requires new high-performance algorithms to be developed in order to give these UAVs the autonomy they require. Similarly, the data-merging functionality offered by Paparazzi must be improved, for several reasons:

– due to low measurement quality, the state of the aircraft needs to be reconstructed from a broad dataset instead of just a single estimate along a single axis;

– parameters previously used in raw form by the control laws of the aircraft need to be usefully filtered and sanity-checked against the estimation model;

– finally, the availability of a data-merging algorithm that can be applied to broader measurement sets can be expected to facilitate the integration of new types of sensor for data merging in the future. For example, vision-based sensors might be able to provide additional redundancy to the system.

Faced with the challenge of achieving safe mini-UAV flight, consolidating our knowledge of the state of an aircraft by means of robust estimation techniques should be viewed as an essential factor in guaranteeing maximum active control of the aircraft and ensuring cell integrity at all times.

2

The State of the Art

2.1. Basic concepts

The mathematical concept of a stochastic process is essential when modeling real dynamic systems where purely deterministic models often prove insufficient. Some physical phenomena that can significantly affect the dynamics of the system remain very poorly understood (such as unmeasured perturbations from noise, mechanical failure, wind, etc.). However, we can still represent them mathematically by stochastic processes. We can integrate various forms of random uncertainty into the global model of the dynamic system by taking advantage of the properties of stochastic processes, often allowing us to establish a relatively general model. In the context of this book, we are only interested in the case of nonlinear state representations that satisfy:

$$\Sigma : \begin{cases} \dot{\mathbf{x}} = f(\mathbf{x}, \mathbf{u}) + \mathbf{v} \\ \mathbf{y} = g(\mathbf{x}, \mathbf{u}) + \mathbf{w} \end{cases} \Leftrightarrow \begin{cases} F(\dot{\mathbf{x}}, \mathbf{x}, \mathbf{u}, \mathbf{v}) = \mathbf{0} \\ G(\mathbf{y}, \mathbf{x}, \mathbf{u}, \mathbf{w}) = \mathbf{0} \end{cases} \quad [2.1]$$

In equation [2.1], $\mathbf{x}$ is the state of the system, $\mathbf{y}$ represents the system outputs, which are assumed to be measurable, and $\mathbf{u}$ denotes a set of known inputs, which are used to control the system, as we shall see with our mini-UAVs later. The parameters $\mathbf{v}$ and $\mathbf{w}$ represent random uncertainties in the evolution and the observations of the system, respectively. Mathematically, we shall represent $\mathbf{v}$ and $\mathbf{w}$ by two continuous time-additive random processes, typically assumed to be independent and uncorrelated. Thus, the state $\mathbf{x}$ of the system, governed by a stochastic algebraic differential equation, is also a random process, and so is the vector of outputs $\mathbf{y}$, which statically depends on the quantity $\mathbf{u}(\mathbf{t})$ and the realizations of $\mathbf{x}(\mathbf{t})$ and $\mathbf{w}(\mathbf{t})$ at any given time $\mathbf{t}$. Hence, the recursive Bayesian estimation problem of the system state, given some set of observables of this state, is to find the probability density $\mathbf{p}(\mathbf{x}(\mathbf{t})|\mathbf{z}_{\mathbf{t_0}:\mathbf{t}'})$, where $\mathbf{t_0}$,$\mathbf{t}$ and $\mathbf{t}'$ are three arbitrary points in time satisfying $\mathbf{t} \geq \mathbf{t_0}$ and $\mathbf{t}' \geq \mathbf{t_0}$. The order relation between $\mathbf{t}$ and $\mathbf{t}'$ allows us to further reduce the estimation problem to:

- a filtering problem whenever $\mathbf{t} = \mathbf{t}'$;
- a prediction problem whenever $\mathbf{t} > \mathbf{t}'$;
- a smoothing problem whenever $\mathbf{t} < \mathbf{t}'$.

Once the problem is solved, the average value of the state can be estimated as follows:

$$\mathbf{E}_{\mathbf{p}(\cdot|\mathbf{z}_{\mathbf{t_0}:\mathbf{t}'})}\left[\mathbf{x}(\mathbf{t})\right] = \int_{\mathbb{R}^{\mathbf{dim}(\mathbf{x}(\mathbf{t}))}} \mathbf{x}(\mathbf{t}) \cdot \mathbf{p}(\mathbf{x}(\mathbf{t})|\mathbf{z}_{\mathbf{t_0}:\mathbf{t}'})\mathbf{dx}(\mathbf{t}).$$

In the case of discrete-time stochastic Markov processes, we can give an outline of the general solution of the estimation problem by considering a general dynamic model of the form:

$$\begin{cases} \mathbf{x}_{\mathbf{t}+1} \sim \mathbf{p}(\mathbf{x}_{\mathbf{t}+1}|\mathbf{x}_\mathbf{t}) \\ \mathbf{y}_\mathbf{t} \sim \mathbf{p}(\mathbf{y}_\mathbf{t}|\mathbf{x}_\mathbf{t}). \end{cases}$$

The probability density of the filtering problem can then be stated as:

$$\mathbf{p}(\mathbf{x}_\mathbf{t}|\mathbf{z}_{\mathbf{t_0}:\mathbf{t}}) = \mathbf{p}(\mathbf{x}_\mathbf{t}|\mathbf{z}_\mathbf{t}, \mathbf{z}_{\mathbf{t_0}:\mathbf{t}-1}) = \frac{\mathbf{p}(\mathbf{z}_\mathbf{t}|\mathbf{x}_\mathbf{t}, \mathbf{z}_{\mathbf{t_0}:\mathbf{t}-1})\mathbf{p}(\mathbf{x}_\mathbf{t}, \mathbf{z}_{\mathbf{t_0}:\mathbf{t}-1})}{\mathbf{p}(\mathbf{z}_\mathbf{t}, \mathbf{z}_{\mathbf{t_0}:\mathbf{t}-1})}.$$

But $\mathbf{p}(\mathcal{A}, \mathcal{B}) = \mathbf{p}(\mathcal{A}|\mathcal{B}) \cdot \mathbf{p}(\mathcal{B})$. Hence:

$$\mathbf{p}(\mathbf{x}_\mathbf{t}|\mathbf{z}_{\mathbf{t_0}:\mathbf{t}}) = \frac{\mathbf{p}(\mathbf{z}_\mathbf{t}|\mathbf{x}_\mathbf{t}, \mathbf{z}_{\mathbf{t_0}:\mathbf{t}-1})\mathbf{p}(\mathbf{x}_\mathbf{t}|\mathbf{z}_{\mathbf{t_0}:\mathbf{t}-1})}{\mathbf{p}(\mathbf{z}_\mathbf{t}|\mathbf{z}_{\mathbf{t_0}:\mathbf{t}-1})} = \frac{\mathbf{p}(\mathbf{z}_\mathbf{t}|\mathbf{x}_\mathbf{t})\mathbf{p}(\mathbf{x}_\mathbf{t}|\mathbf{z}_{\mathbf{t_0}:\mathbf{t}-1})}{\mathbf{p}(\mathbf{z}_\mathbf{t}|\mathbf{z}_{\mathbf{t_0}:\mathbf{t}-1})}.$$

[2.2]

We can also use the formula $p(z_t \mid z_{to:t-1}) = \int_{\mathbb{R}^{\dim(x_t)}} p(z_t \mid x_t)$ $p(x_t \mid z_{to:t-1})dx_t$, which can be derived by integrating the state variable x_t of the following probability density:

$$p(z_t, x_t|z_{to:t-1}) = p(z_t \mid x_t)p(x_t \mid z_{to:t-1}).$$

For the prediction problem, we can derive the required probability density $p(x_t|z_{to:t'})$ by taking the following multidimensional integral of the state variables:

$$p(x_t|z_{to:t'}) = \int_{\mathbb{R}^{(t-t')\dim(x(t))}} p(x_t, x_{t-1}, \cdots, x_{t'+1}, x_{t'}|z_{to:t'})dx_{t':t-1} \quad [2.3]$$

$$= \int_{\mathbb{R}^{(t-t')\dim(x(t))}} \left(\prod_{i=t'+1}^{t} p(x_i|x_{i-1})\right) \cdot p(x_{t'}|z_{to:t'})dx_{t':t-1}. \quad [2.4]$$

Equation [2.4] is known as the Chapman–Komolgorov equation. Finally, the probability density of the smoothing problem can be deduced from the following integral:

$$p(x_t|z_{to:t'}) = \int_{\mathbb{R}^{\dim(x(t))}} p(x_t, x_{t+1}|z_{to:t'})dx_{t+1}$$

$$= \int_{\mathbb{R}^{\dim(x(t))}} p(x_t|x_{t+1}, z_{to:t'}) \cdot p(x_{t+1}|z_{to:t'})dx_{t+1}.$$

We have that:

$$p(x_t|x_{t+1}, z_{to:t'}) = p(x_t|x_{t+1}, z_{to:t}, z_{t+1:t'})$$

$$= \frac{p(z_{t+1:t'}|x_t, x_{t+1}, z_{to:t}) \cdot p(x_t|x_{t+1}, z_{to:t})}{p(z_{t+1:t'}|x_{t+1}, z_{to:t})}$$

$$= p(x_t|x_{t+1}, z_{to:t}).$$

Hence:

$$\begin{aligned}
\mathbf{p}(\mathbf{x_t}|\mathbf{z_{t_0:t'}}) &= \int_{\mathbb{R}^{\dim(\mathbf{x}(\mathbf{t}))}} \mathbf{p}(\mathbf{x_t}|\mathbf{x_{t+1}}, \mathbf{z_{t_0:t'}}) \cdot \mathbf{p}(\mathbf{x_{t+1}}|\mathbf{z_{t_0:t'}})\mathbf{dx_{t+1}} \\
&= \int_{\mathbb{R}^{\dim(\mathbf{x}(\mathbf{t}))}} \mathbf{p}(\mathbf{x_t}|\mathbf{x_{t+1}}, \mathbf{z_{t_0:t}}) \cdot \mathbf{p}(\mathbf{x_{t+1}}|\mathbf{z_{t_0:t'}})\mathbf{dx_{t+1}} \\
&= \int_{\mathbb{R}^{\dim(\mathbf{x}(\mathbf{t}))}} \frac{\mathbf{p}(\mathbf{x_{t+1}}|\mathbf{x_t}, \mathbf{z_{t_0:t}}) \cdot \mathbf{p}(\mathbf{x_t}|\mathbf{z_{t_0:t}})}{\mathbf{p}(\mathbf{x_{t+1}}|\mathbf{z_{t_0:t}})} \\
&\quad \cdot \mathbf{p}(\mathbf{x_{t+1}}|\mathbf{z_{t_0:t'}})\mathbf{dx_{t+1}} \\
&= \mathbf{p}(\mathbf{x_t}|\mathbf{z_{t_0:t}}) \int_{\mathbb{R}^{\dim(\mathbf{x}(\mathbf{t}))}} \frac{\mathbf{p}(\mathbf{x_{t+1}}|\mathbf{x_t}, \mathbf{z_{t_0:t}}) \cdot \mathbf{p}(\mathbf{x_t}|\mathbf{z_{t_0:t'}})}{\mathbf{p}(\mathbf{x_{t+1}}|\mathbf{z_{t_0:t}})}\mathbf{dx_{t+1}}.
\end{aligned}$$

As these results show, in each of the three cases (prediction, filtering, smoothing), computing the probability density involves multidimensional integrals that cannot be simplified to give an analytic solution to the general estimation problem in practice. This complexity is unavoidable except in a few special cases. The best-known example is arguably when the dynamics model is linear and the stochastic variables are normally distributed; in this case, the solution reduces to a Kalman filter.

A wide variety of applications have been successfully handled with this optimal filtering technique, which shows that even this simplified mathematical framework is sufficient to accommodate the majority of estimation problems encountered in real life, and so this special case is perhaps not quite so special as it might seem. Nevertheless, many common estimation problems would benefit from being solved in a more general setting, with less restrictive hypotheses. To do this, we need a more formal solution. A more general solution to the estimation problem outlined above would also expand the range of applications within our grasp, allowing us to tackle more complex problems, including those for which linear methods have proven insufficient. A promising alternative approach to finding an analytic solution to the general problem of estimating the state of a dynamic system is to reinterpret the objectives of the estimation method itself. Instead of working from the probability distribution of the process $\mathbf{x}$, it might be more straightforward to look for a realization of $\mathbf{x}$ directly. From this perspective, given a set known inputs and a set of uncertain measurements, the goal of the estimation algorithm is to construct, over time, an estimate of the vector of the system states $\hat{\mathbf{x}}(\mathbf{t})$, satisfying:

$$\xi : \quad \mathbb{U}_{\mathbf{t}'} \times \mathbb{Y}_{\mathbf{t}'} \longrightarrow \mathbb{R}^{\mathbf{dim}(\mathbf{x}(\mathbf{t}))}$$
$$(\mathbf{u}_{\mathbf{t}_{0:\mathbf{t}'}}, \mathbf{z}_{\mathbf{t}_{0}:\mathbf{t}'}) \longmapsto \hat{\mathbf{x}}(\mathbf{t}) = \xi(\mathbf{u}_{\mathbf{t}_{0:\mathbf{t}'}}, \mathbf{z}_{\mathbf{t}_{0}:\mathbf{t}'}), \hat{\mathbf{x}}(\mathbf{t}) = \underset{\hat{\mathbf{x}}(\mathbf{t}) \in \mathbb{R}^{\mathbf{dim}(\mathbf{x}(\mathbf{t}))}}{\text{argopt}} \quad (\mathbf{J}).$$

This transforms the problem of finding the probability density distribution $\mathbf{p}(\mathbf{x}_{\mathbf{t}} \mid \mathbf{z}_{\mathbf{t}_0:\mathbf{t}})$ into the problem of finding the time realization of the process $\mathbf{x}(\mathbf{t}, \xi)$ that is most likely to explain the observed evolution and variation of the system. However, this alternative approach is only viable if the quality of the estimate can be compared against some independent mathematical criterion $\mathbf{J}$. This characterization of the estimate tells us whether or not the computed time realization is probable, and thus representative of the observed reality. The underlying probabilistic framework of the problem means that we can apply mathematical concepts such as covariance and standard deviation, statistical tests and confidence regions to evaluate the quality of the estimate $\hat{\mathbf{x}}$ over time. These ideas have been successfully implemented many times in the literature. They give a "point by point" approach to estimating the state of the observed dynamic system, the quality of which can be characterized at any given point in time, providing a foundation for a wide variety of other techniques, as we shall see later.

Minimizing the covariance of the estimation error

To tackle the estimation problem from the alternative approach described above, we need to define one or more mathematical criteria for which $\hat{\mathbf{x}}$ is an optimal solution. One initial fairly intuitive idea is to look for a vector estimate of the system states that minimizes the average quadratic distance from the true state $\mathbf{x}$ given the observations $\mathbf{z}_{\mathbf{t}_0:\mathbf{t}'}$, which satisfies:

$$\hat{\mathbf{x}}_{\mathbf{t}} = \underset{\hat{\mathbf{x}} \in \mathbb{R}^{\mathbf{dim}(\mathbf{x}(\mathbf{t}))}}{\text{argmin}} \; \mathbf{E}\{\| \hat{\mathbf{x}}_{\mathbf{t}} - \mathbf{x}_{\mathbf{t}} \|^2 \mid \mathbf{z}_{\mathbf{t}_0:\mathbf{t}'}\}. \qquad [2.5]$$

Solving this optimization problem leads to the estimator known as the Bayesian minimum mean squared error (MMSE) or minimum mean squared deviation (MMSD). The solution can be obtained by developing the criterion in equation [2.5] as follows:

$$\begin{aligned}\mathbf{E}\{\| \hat{\mathbf{x}}_t - \mathbf{x}_t \|^2\} &= \mathbf{E}\{(\hat{\mathbf{x}}_t - \mathbf{x}_t)^{\mathbf{T}}(\hat{\mathbf{x}}_t - \mathbf{x}_t) \mid \mathbf{z}_{t_0:t'}\} \\ &= \mathbf{E}\{(\hat{\mathbf{x}}_t^{\mathbf{T}}\hat{\mathbf{x}}_t) \mid \mathbf{z}_{t_0:t'}\} - 2\hat{\mathbf{x}}_t^{\mathbf{T}}\mathbf{E}\{\mathbf{x}_t \mid \mathbf{z}_{t_0:t'}\} \\ &\quad + \mathbf{E}\{\mathbf{x}_t^{\mathbf{T}}\mathbf{x}_t \mid \mathbf{z}_{t_0:t}\} \\ &= \hat{\mathbf{x}}_t^{\mathbf{T}}\hat{\mathbf{x}}_t - 2\hat{\mathbf{x}}_t^{\mathbf{T}}\mathbf{E}\{\mathbf{x}_t \mid \mathbf{z}_{t_0:t'}\} + \mathbf{E}\{\mathbf{x}_t^{\mathbf{T}}\mathbf{x}_t \mid \mathbf{z}_{t_0:t}\} \\ &= \| \hat{\mathbf{x}}_t - \mathbf{E}\{\mathbf{x}_t \mid \mathbf{z}_{(.)}\} \|^2 + \mathbf{E}\{\| \mathbf{x}_t \|^2 \mid \mathbf{z}_{(.)}\} \\ &\quad - \| \mathbf{E}\{\mathbf{x}_t \mid \mathbf{z}_{(.)}\} \|^2.\end{aligned}$$

This gives:

$$\begin{aligned}\hat{\mathbf{x}}_t &= \underset{\hat{\mathbf{x}} \in \mathbb{R}^{\mathbf{dim}(\mathbf{x}(t))}}{\operatorname{argmin}} \; \mathbf{E}\{\| \hat{\mathbf{x}}_t - \mathbf{x}_t \|^{\mathbf{2}} \mid \mathbf{z}_{t_0:t'}\} \\ &\sim \underset{\hat{\mathbf{x}} \in \mathbb{R}^{\mathbf{dim}(\mathbf{x}(t))}}{\operatorname{argmin}} \; \| \hat{\mathbf{x}}_t - \mathbf{E}\{\mathbf{x}_t \mid \mathbf{z}_{t_0:t'}\} \|^2.\end{aligned}$$

Thus:

$$\hat{\mathbf{x}}_t^{\mathbf{MMSE}} = \mathbf{E}\{\mathbf{x}_t \mid \mathbf{z}_{t_0:t'}\} = \int_{\mathbb{R}^{\mathbf{dim}(\mathbf{x}(t))}} \mathbf{x}(\mathbf{t}) \cdot \mathbf{p}(\mathbf{x}_t | \mathbf{z}_{t_0:t'}) \mathbf{dx}(\mathbf{t}).$$

Thus, the conditional average leads to an estimate that is optimal for any class of quadratic cost functions. It is also unbiased:

$$\begin{aligned}\mathbf{E}\left[\mathbf{x}_t - \hat{\mathbf{x}}_t^{\mathbf{MMSE}}\right] &= \mathbf{E}\left[\mathbf{x}_t\right] - \mathbf{E}\left[\hat{\mathbf{x}}_t^{\mathbf{MMSE}}\right] \\ &= \mathbf{E}\left[\mathbf{x}_t\right] - \underbrace{\mathbf{E}\left[\mathbf{E}[\mathbf{x}_t \mid \mathbf{z}_{t_0:t'}]\right]}_{=\mathbf{E}[\mathbf{x}_t]} = \mathbf{0}.\end{aligned}$$

Hence, in the normally distributed case, knowing the first two conditional moments of $\mathbf{p}(\mathbf{x}_t \mid \mathbf{z}_{t_0:t})$ is necessary and sufficient to solve the optimal filtering problem. These two moments are namely:

– mean $\mathbf{E}\{\mathbf{x}_t \mid \mathbf{z}_{t_0:t}\} = \hat{\mathbf{x}}_t^{\mathbf{MMSE}}$;

– covariance matrix $\mathbf{E}\{(\mathbf{x}_t - \hat{\mathbf{x}}_t^{\mathbf{MMSE}})(\mathbf{x}_t - \hat{\mathbf{x}}_t^{\mathbf{MMSE}})^{\mathbf{T}} \mid \mathbf{z}_{t_0:t}\} = \hat{\mathbf{P}}_t^{\mathbf{MMSE}}$.

The first moment gives the optimal estimate of the problem, and the second moment measures the quality of this estimate. In the normally distributed case, the equations describing the evolution of these first two moments give us the equations of the optimal filter, widely known as the Kalman–Bucy filter.

Maximizing the likelihood of the conditional probability

Another approach can be used to define the optimal estimate of the Bayesian estimation problem. This method still uses conditional probabilities, but moves away from quadratic criteria. The idea is to interpret the state $\mathbf{x_t}$ as an explanatory variable of the system observers. From this perspective, finding an optimal estimate of the state $\mathbf{x_t}$ means finding the vector of parameters that maximizes the likelihood of the observables. This method, developed by Fisher from the 1940s onward, is known as maximum likelihood estimation (MLE). Mathematically, the optimal estimate $\hat{\mathbf{x}}_t^{\mathbf{MLE}}$ is defined as follows:

$$\hat{\mathbf{x}}_t^{\mathbf{MLE}} = \underset{\mathbf{x} \in \mathbb{R}^{\mathrm{dim}(\mathbf{x}(t))}}{\mathrm{argmax}} \; \mathbf{p}(\mathbf{z}_{t_0:t'} \mid \mathbf{x_t}). \qquad [2.6]$$

If the observations are assumed to be independent, equation [2.6] can be rewritten as:

$$\hat{\mathbf{x}}_t^{\mathbf{MLE}} = \underset{\mathbf{x} \in \mathbb{R}^{\mathrm{dim}(\mathbf{x}(t))}}{\mathrm{argmax}} \; \prod_{i=t_0}^{t'} \mathbf{p}(\mathbf{z_i} \mid \mathbf{x_t}). \qquad [2.7]$$

Many formulations of MLE in the literature consider the natural logarithm of the product in equation [2.7]:

$$\hat{\mathbf{x}}_t^{\mathbf{MLE}} = \underset{\mathbf{x} \in \mathbb{R}^{\mathrm{dim}(\mathbf{x}(t))}}{\mathrm{argmax}} \; \sum_{i=t_0}^{t'} \ln(\mathbf{p}(\mathbf{z_i} \mid \mathbf{x_t})).$$

When the elementary conditional probabilities are normally distributed, we can even consider the negative of the natural logarithm of $\mathbf{p}(\mathbf{z}_{t_0:t'} \mid \mathbf{x_t})$:

$$\hat{\mathbf{x}}_t^{\mathbf{MLE}} = \underset{\mathbf{x} \in \mathbb{R}^{\mathrm{dim}(\mathbf{x}(t))}}{\mathrm{argmax}} \; \left(-\ln(\mathbf{p}(\mathbf{z}_{t_0:t'} \mid \mathbf{x_t}))\right).$$

A numerical optimization technique is typically used to compute the estimate at time t. As a result, implementations of the MLE estimator tend to be relatively complex. However, this approach has the advantage of returning an optimal estimate, even when working with large datasets.

Maximizing the a posteriori *likelihood*

Finally, a third perspective can be adopted to define the solution of the estimation problem. This time, we view the optimal estimate of the problem as the argument that maximizes the probability density of the system state given known observations. We can write this as follows:

$$\hat{x}_t = \underset{x \in \mathbb{R}^{\dim(x(t))}}{\operatorname{argmax}} \ p(x_t \mid z_{t_0:t'}). \quad [2.8]$$

Applying Bayes' law to equation [2.8] then gives:

$$\hat{x}_t = \underset{x \in \mathbb{R}^{\dim(x(t))}}{\operatorname{argmax}} \frac{p(z_{t_0:t'} \mid x_t) \cdot p(x_t)}{p(z_{t_0:t'})} \sim \underset{x \in \mathbb{R}^{\dim(x(t))}}{\operatorname{argmax}} \ p(z_{t_0:t'} \mid x_t) \cdot p(x_t).$$

Hence, $\hat{x}_t^{MAP} = \underset{x \in \mathbb{R}^{\dim(x(t))}}{\operatorname{argmax}} \ p(z_{t_0:t'} \mid x_t) \cdot p(x_t)$.

This formulation is known as the maximum *a posterioi* (MAP) estimation. It allows us to incorporate our *a priori* knowledge of the system state into the probability density $p(x_t)$. Formulating hypotheses about $p(x_t)$, regardless of what these hypotheses actually are, then transforms the optimization problem into the problem of finding the state at time t that best explains our observations of the system. This is another example of a "likelihood". It is important to note that:

– if the number of observables $z_{t_0:t'}$ is limited, then the estimate (and its quality) will primarily depend on our *a priori* knowledge of the process $x(t)$ (and the consistency of our representation of this process with reality);

– as the number of observables increases, the MAP estimate weights the information according to its likelihood and our *a priori* knowledge when computing the estimate. Asymptotically, as $t' \longmapsto +\infty$, the MAP and MLE estimators converge to the same value.

As we shall see below, this third approach has been used many times in the literature to design estimation algorithms that can solve a wide variety of problems.

2.2. Literature review

2.2.1. *Important milestones*

The historical origin of estimation theory, which for our purposes is the field that studies the problem of extracting an estimate from noisy data, can be traced as far back as Gauss (18th Century) and Legendre (19th Century). Both helped develop the well-known technique of least-squares approximation, most notably used to find the orbital parameters of the planet Ceres, which was discovered in 1801 by Piazzi. The method of least squares already gave contemporary mathematicians an optimal solution to the problem of estimating unknown parameters from experimental data. It would allow many links to be forged between various technical and applied sciences; the orbit of Ceres in celestial mechanics is just one of many examples. But estimation theory would not progress any further for over a century (see Figure 2.2) until Borel's work on measure theory and its applications to probability theory and Fisher's popularization of the maximum likelihood approach between 1912 and 1922 [FIS 22]. Fisher established an impressive array of results, proofs and demonstrations (Fisher information matrix, convergence, effectiveness of estimation, etc.). This heralded a golden age for the field of estimation, with repeated theoretical breakthroughs over the next 50 years. In the second half of the 1930s, the research into probability theory and stochastic processes performed by Kolmogorov [KOL 22] would allow estimation theory to be completely reimagined along the lines presented in the previous section. Kolmogorov's work, still a cornerstone of modern teaching on probability and stochastic processes, allowed conditional probability (an essential component of estimation theory) to be mathematically defined and formalized based on the Lesbesgue–Radon–Nikodyron theorem.

Kolmogorov also established the foundations of measure theory, laying the groundwork that would allow stochastic processes to be handled in a more appropriately rigorous manner in the context of estimation problems. His theorem for constructing probability measures on infinite-dimensional spaces is one of the most important parts of his legacy. Kolmogorov's axiomatization of probability theory inspired a great deal of further progress in the study of continuous-time stochastic processes, allowing a range of much-needed tools to be developed for contemporary physics. The Chapman–Kolmogorov equation (see section 2.1) is one of the major achievements of this field of research. Later, Wiener would study continuous linear filtering of steady stochastic processes in the 1940s and solve them by introducing the filter equations that bear his name. These equations most notably require solutions of the Wiener–Hopf integral equations in the time domain, which depend on the correlation and spectral density functions of the (noisy) input and target output of the optimal filter; these two parameters are both assumed to be random. After switching to the frequency domain using the bilateral Laplace transform, an analytically known, rational expression can easily be derived for the optimal solution of the estimator filter from the complex spectra of the signals cited above (see Figure 2.1). This was an important step forward, guiding various research projects until the early 1960s, at which point Swerling [SWE 59] and Kalman and Bucy [KAL 60, KAL 61] introduced the state-space techniques of estimation theory that would lead to the definition of the Kalman filter for linear systems. In particular, Kalman showed that, given certain assumptions, the filter thus obtained is optimal in the sense of minimizing the covariance of the estimation error, as well as stable and robust whenever certain controllability and observability conditions are satisfied. The introduction of Kalman filters ignited various lines of theoretical and applied research. The principles established by Kalman allowed numerically efficient and therefore practically useful algorithms to be developed. Data-merging navigation algorithms could now be developed in the aerospace sector, an ideal incubator for this technique due to the vast number of estimation problems encountered. As soon as it was introduced, Kalman filtering was capable of giving answers to several estimation problems in linear settings.

Gradually, from the 1960s onward, the scientific community began to search for methodological solutions to nonlinear estimation problems. Ever better estimation methods capable of solving ever more complex engineering problems became increasingly important. The results obtained by applying linear Kalman filters to models of inherently nonlinear systems were not satisfactory and sometimes even diverged. Over time, the concerted efforts of Stratanovich, Krushner, Zakai, Jazwinski, Gelb, Schweppe and Kailath and more recently, Gordon, Julier and Uhlmann would lead to a selection of nonlinear techniques (such as Extended Kalman Filter (EKF), Unscented Kalman Filter (UKF), or particle filters) capable of solving problems such as that of estimating the state of a dynamic nonlinear system subject to random perturbations and model uncertainties.

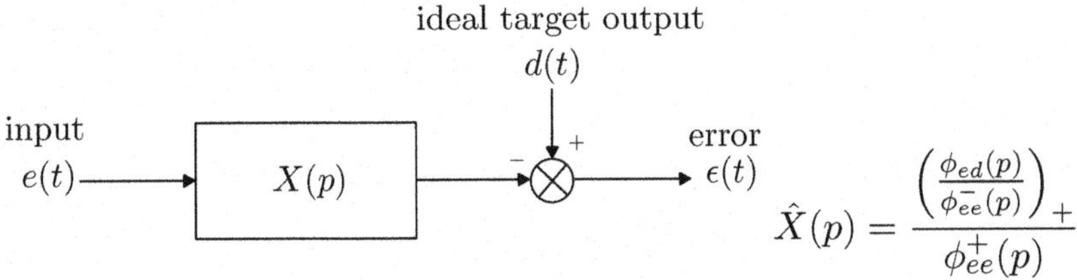

Figure 2.1. *Wiener filter*

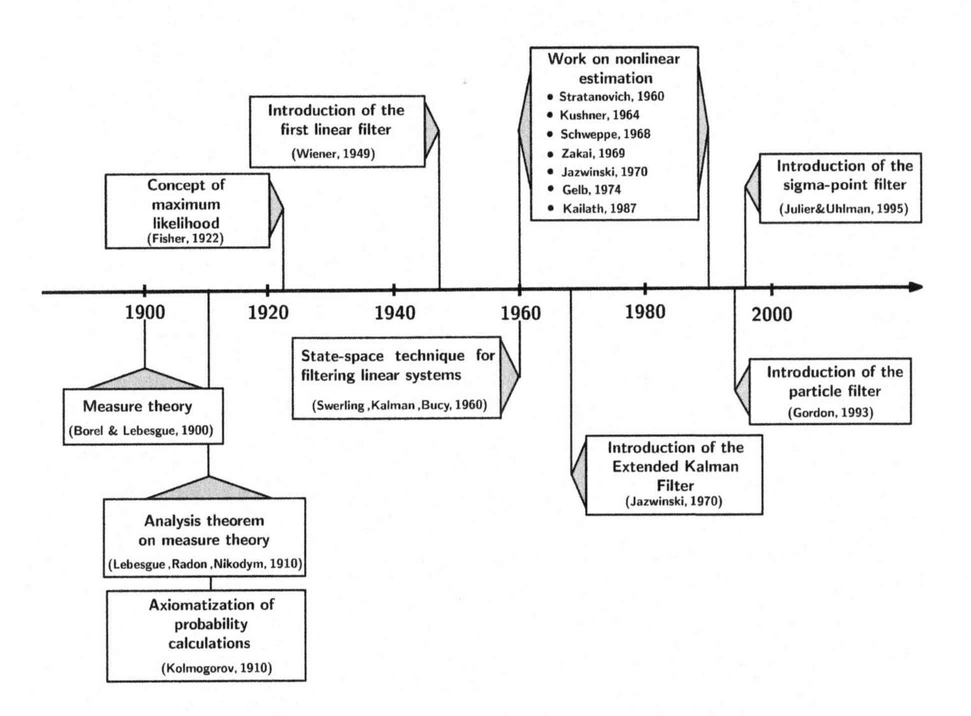

Figure 2.2. *Key milestones in estimation theory*

2.2.2. *Historical overview*

The first-ever state observers were designed in the early 1960s, just as various important theoretical discoveries were unfolding in the field of estimation. Kalman and Bucy [KAL 61] were arguably the first to design an observer based on state-space techniques for linear systems. Their observer produced an optimal and exact result to the recursive formulation of the Bayesian estimation problem. Assuming normally distributed variables, the solution constructed by their observer is optimal in the sense of minimizing the covariance of the estimation error. By analogy with Kalman filters, Viterbi [VIT 67] proposed a variant of the algorithm that could be used with discrete-time Markov processes. The solution constructed by Viterbi's algorithm is optimal in the sense of maximizing the likelihood, an optimality criterion proposed by R.A. Fisher nearly half a century earlier.

The original formulation of the Kalman filter did not explicitly require the system equations to be linear, or the evolution and observation noise parameters to have normally distributed density functions. The only hypothesis stated by Kalman related to the coherence of the random variable (r.v.) required to be conserved when the first- and second-order moments (mean values and covariances) are propagated; in other words, the r.v. was required to be unbiased. However, as it turns out, this hypothesis is never satisfied in nonlinear contexts. This approach was therefore later generalized to the nonlinear case by Breakwell [BRE 67] as the extented Kalman filter by instead applying the principles of the filter to the state model obtained by considering the Taylor expansion of the original nonlinear system around the estimated state. Alternatively, Jazwinski [JAZ 70] proposed a slightly different variant of an EKF by assuming that the second-order terms of the Taylor expansion are bounded. Interest in the linear version of Kalman filtering was initially fanned by the Apollo program. Nonlinear Kalman filters later attracted attention for their implementations in various nonlinear engineering problems [PIC 91, FAR 00, BIJ 08]. The aerospace sector also popularized various algorithms based on the theoretical principles established by Kalman, combined with specific representations such as quaternions to resolve a series of difficulties arising from the use of Euler angles (singularities and discontinuities). Multiplicative Extended Kalman Filters (MEKFs) [MUR 78, LEF 82] are still standard tools in the field today and have been studied by a number of recent research projects. Although techniques inspired by Kalman filtering have been developed extensively by the scientific community, who have adapted and incorporated various other ideas and concepts in order to solve estimation problems such as those involving the H_∞ norm [NAG 91, MAR 93, SMI 95], the convergence properties of these techniques remain limited by the fact that any linearization of the equations of a nonlinear system can never be more than an approximation.

Gradually, the scientific community has begun to adopt mathematical definitions designed to improve stability, a key concern in the context of deterministic observers. This has ultimately led to improvements in the structure of these deterministic observers, rendering them increasingly robust against exogenous perturbations. Luenberger [LUE 71] was the first to formally establish the theoretical principles underlying the observer, which bear his name. His idea was to correct the evolution model after each measurement of the system, expressing this evolution model in a canonical companion-type form known as Brunovsky canonical form. This type of linear observer would later be adopted by almost every approach to designing nonlinear observers. Thau [THA 73] and Kou *et al.* [KOU 75] proposed the first algorithms based on Lyapunov stability analysis, originally developed for linear estimation. This led to the notion of the so-called high-gain observer. Thau gave sufficient conditions for the estimated state to converge to the true state of the system for the class of nonlinear systems whose dynamics can be decomposed into a linear component and a nonlinear component, usually required to satisfy a Lipschitz condition. The term "high-gain" is inspired by the structure of the observer: whenever the nonlinear evolution function has a large Lipschitz constant, even tiny errors between the true state and the estimated state can propagate and be magnified. To counteract the amplification of the error, the gain of the observer must be as high as possible. High-gain observers were later widely used in applications relating to certain classes of biological systems in the mid-1990s.

Since they perform well locally, Kalman filters are robust against noise. As a result, they have become somewhat of a standard in engineering circles. The results obtained by Mehra from the 1970s onward [MEH 70, MOO 75] marked the beginning of adaptive approaches to Kalman filtering. The adaptive aspect of these approaches relates to how they identify the covariance that characterizes the evolution and observation noise terms, which perturb the model of the system [MOO 87]. Moreover, Alpach and Sorenson [ALP 72] developed the first "Gaussian mixture" filter by defining partitions of the state. They proposed the hypothesis that, given any linear system with random inputs characterized by normally distributed probability densities and initial conditions specified as a weighted sum of several normal probability densities, the optimal filter can be computed in finite dimensions. It follows that the *a posteriori* probability density can also be expressed as a weighted sum of Gaussian probability densities. The observer can then be derived by a combined approach as an array of multiple Kalman filters operating in parallel, independently from each other. A number of results about techniques for estimating the attitude in the aerospace sector allowed the estimators used in this field to be simplified. For instance, Higgins [HIG 75] described a simplification of the steady-state regime of a Kalman filter, called a "complementary filter", and Shuster and Oh [SHU 81] were one of the first to construct a filter known as QUEST (QUaternion ESTimation), based on matrices whose orthogonality properties allow a transfer function to be minimized in the sense of least squares. This filter was studied further by the authors [BAR 96, PSI 00, SHU 89, SED 93].

Over time, advancements in our understanding of how to design robust nonlinear observers prompted researchers to employ nonlinear transformation methods to reduce to the more straightforward linear case. Krener and Isidori [KRE 83] were one of the first to study the idea of designing nonlinear observers by applying a change in coordinates (i.e. a diffeomorphism) to the filter equations, thereby reducing to a linear system with an extra injected output. After performing this transformation, the estimated state constructed by a Luenberger observer can be calculated by performing the inverse change in coordinates. This method expresses the original linear system in a canonical observability form. In a similar spirit to Krener's work, Respondek and Krener [RES 85] studied multioutput systems. Various linearization techniques were later developed to support a broader spectrum of nonlinear systems [ZEI 88, BIR 88, KEL 87]. The greatest challenge of these methods is to define a suitable transformation. To overcome this obstacle, Bestle and Zeitz [BES 83] gave necessary and sufficient conditions based on Lie algebras in the case of a single-input/single-output (SISO) system assumed to have a linear observation function. In the late 1980s, other techniques began to emerge, immersing (or embedding) the state in a higher dimensional space in order to circumvent the overly restrictive hypotheses of earlier techniques. Levine and Marino [LEV 86] wrote an article proposing that injective transformations should be found (as opposed to diffeomorphisms, which are bijective), opening the approach to a larger class of nonlinear systems.

After a brief hiatus, intensive research into the design of robust nonlinear observers picked up again in the early 1990s. Among others, Gauthier *et al.* [GAU 92] studied proofs of high-gain properties in order to retrospectively justify the properties of a modified version of an extended Kalman filter inspired by high-gain observers. This research ultimately resulted in a high-gain variant of the extended Kalman filter, combining the rapid convergence of high-gain observers and the strong robustness against noise of extended Kalman filters. Only a decade after high-gain filters were constructed for the first time, Kazantzis and Kravaris [KAZ 98] extended this theory to a class of nonlinear systems with nonlinear real-valued state functions as outputs. More recently, Besancon *et al.* [BES 04] proposed an adaptive high-gain observer for nonlinear systems that depends linearly on unknown parameters. The article [FAR 04] gives a detailed presentation of the design of a high-gain observer for a class of nonlinear multi-input/multi-output (MIMO) systems. Before this, the methods developed for nonlinear systems (high-gain methods and transformation methods) had been either global solutions with very restrictive applicability conditions or local solutions valid only within a specific domain.

Although most research into Kalman filtering aimed to estimate the state of nonlinear dynamic systems with normally distributed noise terms, other (suboptimal) techniques were also developed in the 1990s to eliminate the need for such specific hypotheses. Gordon [GOR 93] proposed a method called particle filtering, based on a so-called sequential Monte Carlo technique for estimating processes. This approach does not make any assumptions about the *a posteriori* probability density, and so can be applied to any stochastic system, even nonlinear systems and/or systems with non-Gaussian distributions. The idea is to approximate the conditional probability density of the filtering problem by a weighted sum of Dirac measurements.

However, the enormous computational cost of this technique led researchers to pursue other paths. Julier *et al.* [JUL 95] introduced the "sigma-point" filtering method based on the idea of approximating the moments of random nonlinear functions. Julier demonstrated the benefit of this technique on a variety of systems inspired by engineering problems: Chapter 7 of [WAN 01] and [CRA 03, DAR 13, SAR 07, XU 08]. We shall present this method in more detail in section 2.5.

The last decade was characterized by the introduction of new techniques that question the classical definition of the estimation error traditionally used by methods of designing nonlinear observers. The first research based on a geometric approach was conducted by various researchers [MAI 05, MAH 08, LAG 08, AGH 02]. Their approach revolves around geometric Lie groups (quaternions, rotation matrices, etc.). Inspired by prior work by Krener and Isidori, it exploits the invariance of some nonlinear systems under certain geometric transformations (rotation, translation, etc.) to construct an observer that performs significantly better than observers defined locally around an equilibrium point. The various results established by Bonnabel and colleagues [BON 07, BON 08, BON 09a, BON 09b] allowed a theory of invariant observers to be developed for a large class of systems with symmetries. New applications in mobile and aerial robotics popularized this approach; in particular, it has proven to be effective for attitude estimation problems [MAR 07, VAS 08, MAR 08, BAR 13]. For example, Bonnabel *et al.* [BON 09b] proposed to use the invariant estimation error in the recursive formulation of an extended Kalman filter to enlarge its domain of convergence *a priori*. This hybrid approach led to the notion of Invariant Extended Kalman Filtering. Other studies conducted by Diop and Fliess [DIO 91] apply numerical differentiation to approximate the noisy signal by polynomial representations, then compute the coefficients of these polynomials algebraically [DIO 09, DIO 12]. Finally, we should mention the work performed by Raïssi *et al.* [RAÏ 10], which constructs observers using interval analysis. Figure 2.3 gives a non-exhaustive classification of the principal methods of stochastic and deterministic estimation.

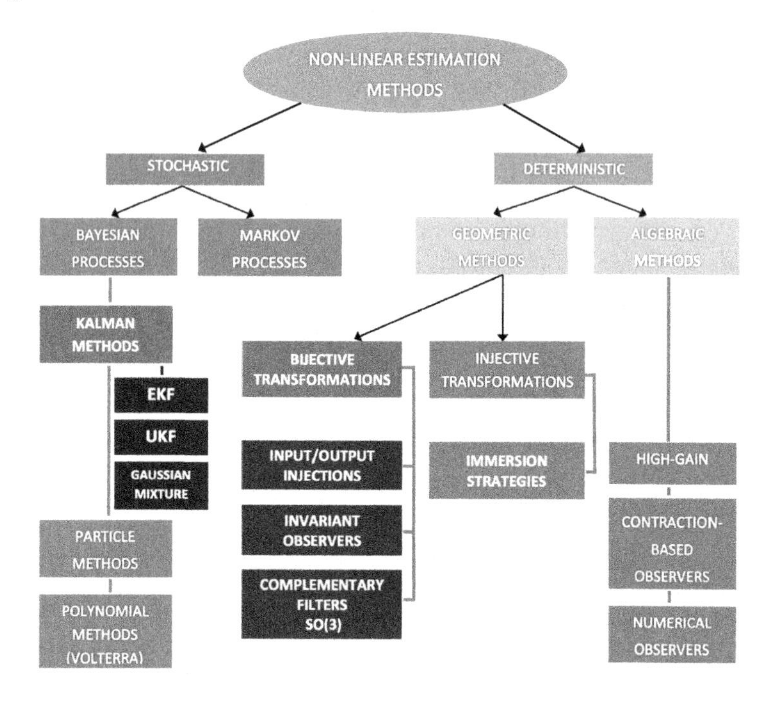

Figure 2.3. *Classification of estimation methods*

2.3. Optimal filtering with linear system models

2.3.1. *Linear Kalman filtering algorithm*

Inspired by the research published before the 1960s, Kalman was one of the first to develop a recursive estimation method and propose algorithms based on state-space techniques for discrete- and continuous-time estimation problems. He was thus able to solve the problem of estimating dynamic system states modeled by linear state representations subject to uncertainty. The solution – the state estimate – is constructed iteratively, and is optimal in the sense of minimizing the covariance of the estimation error. Kalman's optimization applies the least-squares criterion to both continuous and discrete domains. In practice, implementations of Kalman filters are usually constructed discretely, even for continuous systems. This is the case that we shall study below. The ideas used to solve the problem are similar to concepts encountered when studying optimal control. To illustrate the principle of Kalman filtering, consider the simple case of a discrete model of a dynamic system with input parameter $\mathbf{u}$ (assumed to be known), a state parameter $\mathbf{x}$ and an output parameter $\mathbf{y}$. The state and output parameters are, respectively, perturbed by normally distributed white noise terms $\mathbf{w}$ and $\mathbf{v}$ with covariance matrices $\mathbf{Q}$ and $\mathbf{R}$, both centered around zero. The evolution model of the system state $\mathbf{x} \in \mathbb{R}^n$ is usually derived from a model of the underlying physical phenomenon (laws of mechanics, chemistry, electromagnetism, etc.), optionally complemented by an identification procedure whenever meaningful. Finally, the mathematical description of the evolution of the system state is extended by an observation model describing the relation between $\mathbf{x}$ and $\mathbf{y}$. Mathematically, the two models of evolution and observation are represented by the following system of simultaneous equations:

$$\begin{cases} \mathbf{x}_{k+1} = \mathbf{A}_d \cdot \mathbf{x}_k + \mathbf{B}_d \cdot \mathbf{u}_k + \mathbf{w}_k \\ \mathbf{y}_k = \mathbf{H}_d \cdot \mathbf{x}_k + \mathbf{v}_k, \; k \in \mathbb{N} \end{cases}, \qquad [2.9]$$

where:

$$E[\mathbf{w}_i \cdot \mathbf{w}_j^T] = \mathbf{Q}_i \cdot \delta_{ij}, \; E[\mathbf{v}_i \cdot \mathbf{v}_j^T] = \mathbf{R}_i \cdot \delta_{ij},$$

$$E[\mathbf{v}_i \cdot \mathbf{w}_j^T] = 0, E[\mathbf{v}_i] = 0, E[\mathbf{w}_j] = 0, \quad \forall i, j.$$

In equation [2.9], δ_{ij} denotes the Kronecker delta. In the case of a continuous system with deterministic input **u** and state **x** sampled at a rate T:

$$\begin{aligned}
\mathbf{A}_d &= e^{A \cdot T}, \\
\mathbf{B}_d &= \int_0^T B \cdot e^{A \cdot \nu} d\nu, \\
\mathbf{H}_d &= H, \ \mathbf{R}_d = \frac{R}{T}, \ \mathbf{Q}_d = \int_0^T e^{A \cdot \nu} \cdot Q \cdot e^{A^T \cdot \nu} d\nu.
\end{aligned} \qquad [2.10]$$

The matrices A, B, H, Q, and R in equation [2.10] are assumed to be time dependent. The equations of the Kalman filter estimator then invoke certain fundamental properties intrinsically satisfied by Gaussian stochastic processes in order to derive an explicit solution to equations [2.2] and [2.4]. Specifically:

1) every affine transformation of a Gaussian r.v. is also a Gaussian r.v.;

2) the characteristic function is also a Gaussian. Moreover, the sum of any quadratic forms is a quadratic form, and the product of any exponentials of quadratic forms is an exponential of a quadratic form. This means that the Gaussian structure is preserved under multiplication and convolution;

3) finally, any Gaussian probability density function is fully characterized by its first two moments, i.e. the mean and the variance.

Applying these properties allows us to reformulate the recursive parts of equations [2.2] and [2.4] as recursive equations in the mean and the covariance of each probability density. Whenever a new measurement becomes available, the state is updated by a specific prediction/correction scheme. Therefore, writing $\hat{\mathbf{x}}_{i|j}$ for the estimate of the state **x** at time iT given the set of all measurements performed before time jT, denoted as $\mathbf{z}^j = [\mathbf{z}_1, \cdots, \mathbf{z}_j]$, and defining $\mathbf{P}_{i|j}$ as the covariance matrix of the estimation error:

– from a known estimate $\hat{\mathbf{x}}_{k|k}$, the filter begins by predicting the future state of the system, from the model, denoted here by $\hat{\mathbf{x}}_{k+1|k}$. The covariance matrix of the prediction error is denoted $\mathbf{P}_{k+1|k}$ and defined as follows:

$$
Prediction \begin{cases} \hat{\mathbf{x}}_{k+1|k} &= E[\mathbf{A}_d \cdot \hat{\mathbf{x}}_k + \mathbf{B}_d \cdot \mathbf{u}_k + \mathbf{w}_k \mid \mathbf{z}^k] \\ &= \mathbf{A}_d \cdot \hat{\mathbf{x}}_{k|k} + \mathbf{B}_d \cdot \mathbf{u}_k \\ \mathbf{P}_{k+1|k} &= E[(\mathbf{x}_{k+1} - \hat{\mathbf{x}}_{k+1}) \cdot (\mathbf{x}_{k+1} - \hat{\mathbf{x}}_{k+1})^T \mid \mathbf{z}^k] \\ &= E[(\mathbf{x}_{k+1} - \hat{\mathbf{x}}_{k+1|k}) \cdot (\mathbf{x}_{k+1} - \hat{\mathbf{x}}_{k+1|k})^T] \\ &= \mathbf{A}_d \cdot \mathbf{P}_{k|k} \cdot \mathbf{A}_d^T + \mathbf{Q}_d \\ \hat{\mathbf{y}}_{k+1|k} &= \mathbf{H}_d \cdot \hat{\mathbf{x}}_{k+1|k} \end{cases} \qquad [2.11]
$$

In equation [2.11], the predicted covariance matrix $\mathbf{P}_{k+1|k}$ is independent of the system observations $\mathbf{z}^k$, which allows us to compute the mean squared error committed *a priori* by the filtering operation.

– The predicted state $\hat{\mathbf{x}}_{k+1|k}$ is then corrected using the measurements $\mathbf{z}_{k+1}$ performed at time $(k+1)T$ to obtain the estimated state $\hat{\mathbf{x}}_{k+1|k+1}$. This correction, which is linear in its measurements, is weighted by the gain $\mathbf{K}_{k+1}$ in such a way as to minimize the covariance of the estimation error. The weights are calculated from the covariance matrix of the error between the raw measurements $\mathbf{z}$ and the predicted outputs $\hat{\mathbf{y}}_{k+1|k}$, which is denoted $\mathbf{P}_{\mathbf{zz},k+1|k}$, and from the so-called cross-covariance matrix of the error between the output and the state, denoted $\mathbf{P}_{\mathbf{xz},k+1|k}$. The correction gain terms $\mathbf{K}_{k+1}$ are also calculated offline from the values of $\mathbf{P}_{k+1|k}$ and $\mathbf{H}_d$, but independently of the realizations of the process $\mathbf{z}$. If the system dynamics are trivial ($\mathbf{Q}_d = 0$), e.g. when estimating constant parameters, the correction gain always tends to 0, as does the covariance of the filtering error. When estimating constant parameters, this means that new observations are no longer necessary for the correction step after a certain period of time has elapsed.

$$
Update \begin{cases} \hat{\mathbf{x}}_{k+1|k+1} &= \hat{\mathbf{x}}_{k+1|k} + \mathbf{K}_{k+1} \cdot (\mathbf{z}_{k+1} - \hat{\mathbf{y}}_{k+1|k}) \\ \mathbf{K}_{k+1} &= \mathbf{P}_{\mathbf{xz},k+1|k} \cdot \mathbf{P}_{\mathbf{zz},k+1|k}^{-1} \\ \mathbf{P}_{k+1|k+1} &= E[(\mathbf{x}_{k+1} - \hat{\mathbf{x}}_{k+1}) \cdot (\mathbf{x}_{k+1} - \hat{\mathbf{x}}_{k+1})^T \mid \mathbf{z}^{k+1}] \\ &= E[(\mathbf{x}_{k+1} - \hat{\mathbf{x}}_{k+1|k+1}) \cdot (\mathbf{x}_{k+1} - \hat{\mathbf{x}}_{k+1|k+1})^T] \\ &= \mathbf{P}_{k+1|k} - \mathbf{K}_{k+1} \cdot \mathbf{P}_{\mathbf{zz},k+1|k} \cdot \mathbf{K}_{k+1}^T \\ \mathbf{P}_{\mathbf{xz},k+1|k} &= E[(\mathbf{x}_{k+1} - \hat{\mathbf{x}}_{k+1}) \cdot (\mathbf{z}_{k+1} - \hat{\mathbf{z}}_{k+1})^T] \\ &= \mathbf{P}_{k+1|k} \cdot \mathbf{H}_d^T \\ \mathbf{P}_{\mathbf{zz},k+1|k} &= E[(\mathbf{z}_{k+1} - \hat{\mathbf{z}}_{k+1}) \cdot (\mathbf{z}_{k+1} - \hat{\mathbf{z}}_{k+1})^T] \\ &= \mathbf{H}_d \cdot \mathbf{P}_{k+1|k} \cdot \mathbf{H}_d^T + \mathbf{R}_d \end{cases} \qquad [2.12]
$$

The algorithm is initialized by a state $\hat{\mathbf{x}}_{0|0}$ that is known or assumed arbitrary and a covariance matrix $\mathbf{P}_{0|0}$ that is quantified accordingly.

2.3.2. *Extension to nonlinear models*

With the same notation as above, we shall now derive the Kalman filter equations that allow us to estimate non-measurable states of a system modeled by a discrete nonlinear state representation. Our approach to tackling this problem using the results from the linear case unfolds in two steps. First, we describe the real system mathematically by a nonlinear function that represents the dynamics of the state,

$$\mathbf{x}_{k+1} = f(\mathbf{x}_k, \mathbf{u}_k, \mathbf{w}_k, k), \qquad [2.13]$$

as well as a set of nonlinear statistical relations modeling the observations performed on the real system, which might therefore include information relating to the sensors, together with their imperfections and/or dynamics. The noise terms in the state and the measurement noise terms, taken as the inputs of the model, are again assumed to be normally distributed, centered and uncorrelated. Another measurement noise term, assumed to be additive, is also incorporated into this observation model, as follows:

$$\mathbf{z}_k = h(\mathbf{x}_k, \mathbf{u}_k, \mathbf{k}) + \mathbf{v}_k. \qquad [2.14]$$

At this point, the nonlinear nature of the estimation problem becomes a major obstacle. After applying nonlinear functions f and h, we can no longer guarantee that the noise terms continue to be normally distributed as they are propagated through the model. The properties outlined above in the linear case are no longer satisfied. Although linear Kalman filtering can work with colored noise for which a Markov representation is known, it cannot yield a solution to the nonlinear problem stated above. Extended Kalman filtering instead produces a suboptimal solution to the problem by performing a dual linearization of the relations associated with f and h with respect to $\hat{\mathbf{x}}$. There are several ways to extend traditional Kalman filters to estimate the state of a discrete nonlinear system by suboptimally minimizing criteria phrased in terms of the covariance of the estimation error. In each case, the distributions of the errors in the state $\mathbf{x}$ and the measurements $\mathbf{y}$ are approximated by

Gaussian distributions, which once again allows linear Kalman filters to be applied.

2.4. Approximating the optimal filter by linearization: the EKF

The method jointly developed by Breakwell and Jazwinski relies heavily on the filtering techniques developed by Kalman, adapting them so they can be applied to nonlinear systems. This method continues to use the same prediction/correction scheme as Kalman, but applies it to a new idea proposed by Breakwell. Namely, he said that the problem of estimating the state of a nonlinear system subject to stochastic perturbations can be approximated up to a certain accuracy by a linear Kalman filter based on the tangent linear model of the nonlinear system, computed around the estimated trajectory. Indeed, provided that the nonlinearities in the evolution process [2.13] and the observation process [2.14] are not excessively large and the variance of the noise terms is not too high, the true state $\mathbf{x}_k$ of the nonlinear system can be estimated locally from the linearized model with reasonable confidence in its validity. Thus, the estimated state is constructed by combining a prediction based directly on the nonlinear equations of the system and a correction that essentially follows the same principles as the linear Kalman filter in which the matrices $\mathbf{A}_d$ and $\mathbf{H}_d$ are now associated with the tangent linear model of the nonlinear system. The hope is that $\hat{\mathbf{x}}_k$ will represent a good estimate of $\mathbf{x}_\mathrm{k}$. Consequently, we will be able to locally approximate the probability densities of Kalman's original estimation errors with Gaussian probability densities by assuming that the prediction error

$$\epsilon_{k+1|k} = \mathbf{x}_{k+1} - \hat{\mathbf{x}}_{k+1|k} = f(\mathbf{x}_k, \mathbf{u}_k, \mathbf{w}_k, k) - f(\hat{\mathbf{x}}_{k|k}, \mathbf{u}_k, k) \qquad [2.15]$$

can be derived in a reasonably representative form from the Taylor expansion at $\hat{\mathbf{x}}_{k|k}$, truncated to first or second order. There is a very natural choice of point $\hat{\mathbf{x}}_k$ around which to linearize the model. Assuming that contributions from higher order terms are negligible and given that the function $f(\mathbf{x}_k)$ is used in the prediction stage, it seems reasonable to linearize this function in the neighborhood of the most recent estimated state $\hat{\mathbf{x}}_{k|k}$, since the latter is corrected and updated at every computational step. This point is also the best possible known estimate of $\hat{\mathbf{x}}_k$. Thus, to the first order, equation [2.15] becomes:

$$\epsilon_{k+1|k} \approx (\nabla f)_{\mathbf{x}=\hat{\mathbf{x}}_{k|k}} \epsilon_{k|k},$$

where $(\nabla f)_{\mathbf{x}=\hat{\mathbf{x}}_{k|k}}$ is the Jacobian matrix of f at $\mathbf{x} = \hat{\mathbf{x}}_{k|k}$. Recall that:

$$\left[(\nabla f)_{\mathbf{x}=\hat{\mathbf{x}}_{k|k}}\right]_{i,j} = \left.\frac{\partial f_i}{\partial \mathbf{x}_j}\right|_{\mathbf{x}=\hat{\mathbf{x}}_{k|k}}.$$

The mathematical formulations of linear Kalman filters and first-order extended Kalman filters are very similar: the matrices $\mathbf{A}_d$ in the prediction step and $\mathbf{H}_d$ in the correction step are simply replaced by the Jacobian matrices $(\nabla f)_{\mathbf{x}=\hat{\mathbf{x}}_{k|k}}$ and $(\nabla h)_{\mathbf{x}=\hat{\mathbf{x}}_{k+1|k}}$. However, to implement the EKF, the correction gain, and therefore the covariance matrices, must be computed online. Indeed, in the case of a discrete nonlinear system, these two Jacobian matrices depend on the estimates $\hat{\mathbf{x}}_{k|k}$ and $\hat{\mathbf{x}}_{k+1|k}$. Thus, there are several conceivable variants of this algorithm as soon as a linear approximation is performed. For example, since one of the underlying principles of the EKF is to linearize the observation model $h(\mathbf{x}_k)$ in the neighborhood of the prediction $\hat{\mathbf{x}}_{k+1|k}$, the filtered value $\hat{\mathbf{x}}_k$ should be closer to the true state of the system than $\hat{\mathbf{x}}_{k+1|k}$. We could therefore relinearize this model around this new point, hoping to reduce the error, and recalculate the correction. This process can be iterated until the variation in the estimate produced by each estimator is no longer significant. By an analogous procedure, the accuracy of the prediction step can also be improved by recalculating the point around which the evolution model $f(\mathbf{x}_k)$ is linearized. This technique, known as iterative extended Kalman filtering, improves the performance of the estimation process whenever there are rapidly changing nonlinearities. Bell [BEL 94] showed that this iterative formulation of the equations for correcting the state is equivalent to maximizing the likelihood (see equation [2.6]) of the correction term of the predicted state, which can be accomplished using a Gauss–Newton method. However, EKF does have certain limitations. For instance, despite (or perhaps because of) its simplicity, it is important to note that:

– linear or quadratic approximations of the evolution equations [2.13] and observation equations [2.14] can only yield good results so long as these transformations remain valid. Therefore, this filter can only be used with estimation problems with weakly nonlinear system dynamics (to preserve the validity of the tangent linear model) and perturbations with limited variance;

– furthermore, computing the Jacobian matrices often leads to numerical errors that compound any other modeling or hypothesis-related errors.

To improve the robustness of the estimation in terms of its convergence and the numerical accuracy of the filter, we need to turn to more recent algorithms that take advantage of a combination of matrix factorizations and finite centered difference approximations of the Jacobian matrices (∇f) and (∇h). These techniques yield a new formulation of the Kalman filter that is similar to the EKF but which does not require any derivatives to be computed numerically to find the covariance matrix of the estimation errors. One of these techniques is based on a square-root-type reformulation of the equations of the EKF, popular for its good numerical stability.

2.5. Approximating the optimal filter by discretization: the Sigma-Points Kalman Filter

2.5.1. *Prediction and sigma points*

The methodological work that Julier and Uhlmann conducted together after 1995 led them to develop an original technique for estimating the state of nonlinear dynamic systems subject to random perturbations. Their idea was to generate a set of points that is representative of the statistical properties of the error committed by estimating the system state. Coupled with the recursive equations of the Kalman filter, this alternative approach to estimation by sampling the state space can be used to dynamically construct an estimate of the system state (first-order moment or conditional expectation) and the error (second-order moment or covariance of the error) as the moments of this set of points, which are updated over time. The weights of each sample point in the computation are defined *a priori*. Unlike Jazwinski's approach, which estimates the mean of the estimated state and the covariance of the estimation error iteratively from a linearization of the model equations, this method is motivated by the intuition that *it is easier to approximate a distribution [...] than a nonlinear function*. In other words, instead of considering an approximation of the nonlinear function, we should instead attempt to approximate the first- and second-order moments of $\mathbf{x}_k$ and $\mathbf{x}_k - \hat{\mathbf{x}}_k$, respectively.

The prediction step aims to calculate the future state $\hat{\mathbf{x}}_{k+1|k}$ and the covariance matrix $\mathbf{P}_{k+1|k}$ from $\mathbf{P}_{k|k}$ using a nonlinear reformulation of equation [2.11]. To do this, we choose a set of $p+1$ vectors from the state space, denoted $\boldsymbol{\mathcal{X}}_{k|k}^{(i)}, i \in [\![0; p]\!], p \in \mathbb{N}^+$, and called the *sigma points*, in such a way that their average, used to find the estimate and the covariance of the estimation error, can be deduced from the following relations:

$$\sum_{i=0}^{p} \mathbf{W}^{(i)} \boldsymbol{\mathcal{X}}_{k|k}^{(i)} = \hat{\mathbf{x}}_{k|k} \quad \text{and}$$

$$\sum_{i=0}^{p} \mathbf{W}^{(i)} \left(\boldsymbol{\mathcal{X}}_{k|k}^{(i)} - \hat{\mathbf{x}}_{k|k} \right) \left(\boldsymbol{\mathcal{X}}_{k|k}^{(i)} - \hat{\mathbf{x}}_{k|k} \right)^T = \mathbf{P}_{k|k}.$$

The samples of the state space and the weighted average calculation are designed to capture certain properties (mean values and covariance) of the state and error distributions. Each of the $p+1$ weights $\mathbf{W}^i$ can be either positive or negative. To ensure that the estimate is unbiased, they must satisfy the following relation:

$$\sum_{i=0}^{p} \mathbf{W}^{(i)} = 1. \qquad [2.16]$$

The key idea of the method is then to apply the nonlinear transformation f to each of the $p+1$ sigma points to obtain a cloud of updated elements. The statistical properties of this image set give estimates of the images of the mean state and the covariance of the error after applying the nonlinear function f. Thus, in the prediction step, we need to compute the mean values and covariances of the $p+1$ images of the sigma points under f. The calculations of this step are summarized below. Figure 2.4 also illustrates the underlying idea of the method developed by Julier and Uhlmann.

2.5.2. *Interpretation of the sigma-point approach*

We can interpret and understand the underlying principles of sigma-point sampling a little better by considering an analysis of the Sigma-Points Kalman Filter (SPKF) algorithm implicitly based on a technique known as

Weighted Statistical Linear Regression (WSLR) [LEF 02]. At its core, this technique attempts to find a parametric cause-and-effect relation (a regression) that explains the available observations or samples as accurately as possible. The relation (or measurements) can be either linear or nonlinear. It is typically obtained by minimizing a least-squares criterion phrased in terms of p reference datapoints. The WSLR method is unique in that it weights each quadratic distance term when optimizing the least-squares criterion, then performs the optimization with the regression parameters as the decision variables. If the reference data are representative of an r.v., then the result obtained by the WSLR is equivalent to a linear or nonlinear model of this r.v. In the case of an affine model, the technique is known as stochastic linearization. From a purely statistical perspective, the result gives a better image of the estimate of the stochastic process than we would, for example, obtain by approximating with a Taylor expansion up to a certain order. Proposition 2.1 summarizes the theoretical principles of the WSLR when finding an affine model based on reference data that correspond to the images of the sigma points in the standard UKF algorithm.

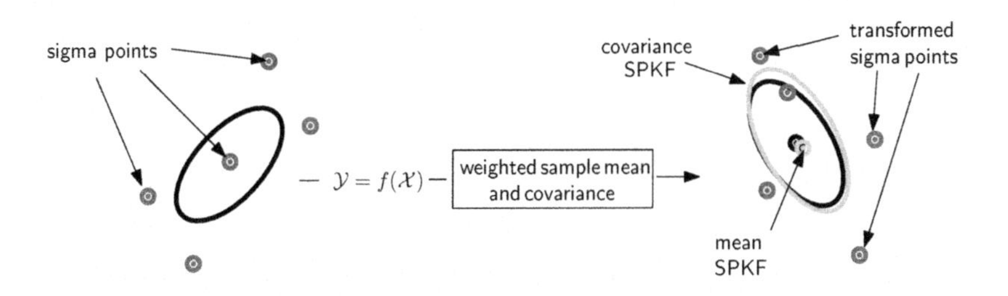

Figure 2.4. *Principle of the sigma-point approach*

PROPOSITION 2.1.– Consider the set of $p+1$ sigma points introduced earlier: $\left\{\boldsymbol{\mathcal{X}}_{k|k}^{(i)}/i \in [\![0;p]\!]\right\}$, each defined at some time $k \in \mathbb{N}$.

Consider also the nonlinear function $\mathbf{g}$.

In the following, write $\gamma_{k+1|k}^{i}$ for the image under $\mathbf{g}$ of the ith sigma point $\boldsymbol{\mathcal{X}}_{k|k}^{(i)}$. We have that:

$$\forall i \in [\![0;p]\!], \gamma_{k+1|k}^{i} = \mathbf{g}(\boldsymbol{\mathcal{X}}_{k|k}^{(i)}).$$

To find the linear model that gives the best explanation of the action of the function $\mathbf{g}$ at each sigma point $\boldsymbol{\mathcal{X}}_{k|k}^{(i)}$, we minimize the following weighted quadratic criterion J:

$$J = \sum_{i=0}^{p} \mathbf{W}^{(i)} (\epsilon_k^{(i)})^T \epsilon_k^{(i)},$$

where

$$\begin{aligned} \forall i \in [\![0;p]\!],\ \epsilon_k^{(i)} &= \gamma_{k+1|k}^{(i)} - (\mathbf{A}\boldsymbol{\mathcal{X}}_{k|k}^{(i)} + \mathbf{b}) \\ &= \mathbf{g}(\boldsymbol{\mathcal{X}}_{k|k}^{(i)}) - (\mathbf{A}\boldsymbol{\mathcal{X}}_{k|k}^{(i)} + \mathbf{b}). \end{aligned}$$

The solution satisfies:

$$(\mathbf{A}^*, \mathbf{b}^*) = \underset{(\mathbf{A},\mathbf{b})}{\text{argmin}}\ J.$$

REMARK 2.1.– In some cases, it may make sense to add the following constraint to the WSLR method:

$$\sum_{i=0}^{p} \mathbf{W}^{(i)} = 1.$$

If these sigma points are intended to represent a stochastic process at a certain moment in time, the problem is equivalent to computing a statistical linear regression that preserves the distribution of the state $\mathbf{x}$ whenever the regression points $\boldsymbol{\mathcal{X}}_{k|k}^{(i)}$ are chosen to accurately reproduce the mean and covariance of the distribution, with the implicit hypothesis that the stochastic process is effectively normally distributed. When applying this approach to the system state estimation problem, it has been shown that, with the least-squares criterion, the solution is obtained when $\mathbf{A} = \hat{\mathbf{P}}_{\mathbf{xy}}^T \mathbf{P}^{-1}$ and $\mathbf{b} = \hat{\mathbf{y}} - \mathbf{A}\hat{\mathbf{x}}$.

This method, also known as stochastic linearization, demonstrates the efficiency of sigma-point filtering compared to estimation algorithms based on classical linearization techniques involving first-order Taylor expansions of the nonlinear state model. Figure 2.5 gives a brief overview of each implicit step in the sigma-points approach, compared to the corresponding step of the EKF algorithm. In the figure, an arbitrary transformation represented by a nonlinear function $\mathbf{g}$, plotted as $\mathbf{y} = \mathbf{g}(\mathbf{x})$, is applied to the distribution of the Gaussian r.v. $\mathbf{x}$, shown in dark green on the horizontal axis. The images $\gamma^{(i)}$ of the sigma points are shown in yellow on the vertical axis. As noted above, the first step in approximating the function $\mathbf{g}$ at $\hat{\mathbf{x}}$ is to truncate its Taylor expansion at $\hat{\mathbf{x}}$ to first order (red dotted line), in the same way as the EKF technique. Therefore, the mean and covariance (in red) are propagated through this initial linear approximation of the function $\mathbf{g}(\mathbf{x})$. As a result, the first two moments of the image of the r.v. are inaccurate relative to the true quantities, shown here in light green on the vertical axis. By contrast, the WSLR technique propagates the sigma points, which were deliberately chosen to be representative of the mean $\boldsymbol{\mathcal{X}}^{(0)} = \hat{\mathbf{x}}$ and the bounds $\boldsymbol{\mathcal{X}}^{(1)} = \hat{\mathbf{x}} - \sigma_{\mathbf{x}}$ and $\boldsymbol{\mathcal{X}}^{(2)} = \hat{\mathbf{x}} + \sigma_{\mathbf{x}}$, where $\sigma_{\mathbf{x}}$ is the standard deviation of the distribution of $\mathbf{x}$. Propagating these points yields the images of the sigma points γ_i (also known as the *a posteriori* sigma points), shown in yellow on the $\mathbf{y}$-axis. We can already see that the line $y_{sp} = \mathbf{Ax} + \mathbf{b}$ obtained by WSLR, which passes near the points with coordinates $(\hat{\mathbf{x}}, g(\hat{\mathbf{x}})), (\boldsymbol{\mathcal{X}}^{(1)}, \gamma^{(1)})$ and $(\boldsymbol{\mathcal{X}}^{(2)}, \gamma^{(2)})$, ultimately reflects the true values of the first two moments of the r.v., whereas the linearization used by EKF does not. The result, shown in purple, is a clear improvement over the approximation derived from the first-order Taylor expansion.

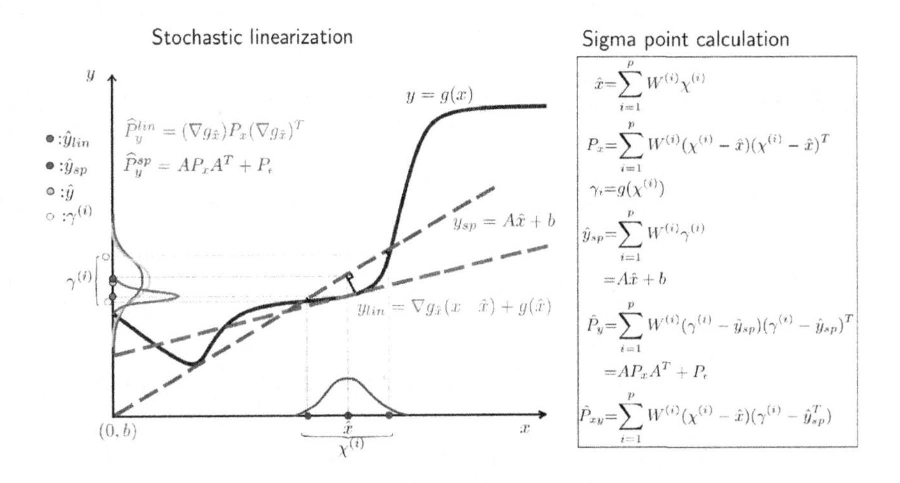

Figure 2.5. *Comparison of the sigma-point and EKF approaches. For a color version of this figure, see www.iste.co.uk/condomines/kalman.zip*

2.5.3. *Scaled UKF*

The method used to select the set of $p+1$ sigma points $\boldsymbol{\mathcal{X}}_{k|k}^{(i)}$ is essential, given that it will be repeated at every sampled time kT. The objective of this method is to find samples of the state space whose statistical mean and covariance are maximally representative of the mean and covariance of the distribution of $\mathbf{x}_k$. Depending on how these points are defined, higher order moments of the distribution of $\mathbf{x}_k$ can also be accurately represented. Thus, the accuracy of the predicted values for the state $\hat{\mathbf{x}}_{k+1|k}$ and the covariance of the estimation error $\mathbf{P}_{k+1|k}$ depends on how the sigma points are constructed. One straightforward idea is to choose the sigma points to be symmetric around some central point [JUL 04, JUL 00]. If $n = dim(\mathbf{x})$, then, by choosing $2n + 1$ symmetric points around $\hat{\mathbf{x}}_{k|k}$ that satisfy $\hat{\mathbf{x}}_{k+1|k} = \sum_{i=0}^{p} \mathbf{W}^{(i)} \boldsymbol{\mathcal{X}}_{k+1|k}^{(i)}$ and giving them identical weights, the mean, covariance and every centered moment of odd order calculated from these points will be identical to the characteristic parameters of the distribution of the true state $\mathbf{x}_k$. However, this choice cannot exactly reproduce the centered moments of the distribution of $\mathbf{x}_k$ that have even orders greater than or equal to four. To overcome this problem, we can view the first weight $\mathbf{W}^{(0)}$ as an adjustable parameter that can be varied to minimize the approximation errors induced in the centered moments of the distribution of $\mathbf{x}_k$ with order strictly greater than three. Another approach is to construct the smallest possible set of sigma points [JUL 02]. The complexity of the UKF algorithm directly depends on the number of sigma points considered at each time kT. If the model involves a high number of states and a real-time implementation is required, it can be useful to find the minimum number of sigma points that guarantee that the distribution of $\mathbf{x}_k$ is sufficiently well represented. As shown by Julier in [JUL 03], in an n-dimensional space, $n+1$ affinely independent sigma points (which define a *simplex* in the space) are necessary and sufficient to capture the average and the spread. Thus, given certain hypotheses on the state and the covariance of the initial state, $n+1$ is the minimum number of sigma points needed at each moment kT. Any given method for constructing and selecting the number of sigma points must strike a compromise between: *the computational complexity of the estimation algorithm*, which, as noted above, strongly depends on the number of sigma points considered at each time step kT; *the quality of the estimate constructed by the method*, which depends on how well the sigma points capture the spread of the distribution over the state space; and *the representativeness of the sample* relative to the original distribution that characterizes $\mathbf{x}_k$, especially for the moments of order greater than two, which influences the accuracy of the mean and covariance of the predicted values. The method known as Scaled Unscented Kalman Filtering proposes an effective way of simultaneously meeting all of these criteria when generating the sigma points. The idea is to select a set of $2n + 1$ points $\mathcal{X}_{k|k}^{(i)}$ and match them with the weights $\mathbf{W}^{(i)}$ listed in Table 2.1. For simplicity, the notation $\mathbf{A} \pm \mathbf{u}$ denotes the operation of adding the vector u to each column of the matrix $\mathbf{A}$, and $\sqrt{\mathbf{P}_{\mathbf{k}|\mathbf{k}}}$ denotes the square root matrix of the covariance of the error returned by Choleksy factorization [WAN 00, LAV 04].

Samples (*sigma points*) $i \in \{1, \cdots, n\}$	Weights for computing the mean $i \in \{1, \cdots, n\}$	Weight for computing the covariance $i \in \{1, \cdots, n\}$
$\boldsymbol{\mathcal{X}}_{k\|k}^{(0)} = \hat{\mathbf{x}}_{k\|k}$ $\boldsymbol{\mathcal{X}}_{k\|k}^{(i)} = \hat{\mathbf{x}}_{k\|k} + \Delta_{k\|k}(i)$ $\boldsymbol{\mathcal{X}}_{k\|k}^{(i)} = \hat{\mathbf{x}}_{k\|k} - \Delta_{k\|k}(i)$	$\mathbf{W}_{(m)}^{(0)} = \frac{\lambda}{n+\lambda}$ $\mathbf{W}_{(m)}^{(i)} = \frac{1}{2(n+\lambda)}$ $\mathbf{W}_{(m)}^{(i+n)} = \frac{1}{2(n+\lambda)}$	$\mathbf{W}_{(c)}^{(0)} = \frac{\lambda}{n+\lambda} + (1 - \alpha^2 + \beta)$ $\mathbf{W}_{(c)}^{(i)} = \frac{1}{2(n+\lambda)}$ $\mathbf{W}_{(c)}^{(i+n)} = \frac{1}{2(n+\lambda)}$
$\Delta_{k\|k} = \sqrt{(n + \lambda)P_{k\|k}}$	$\lambda = \alpha^2(n + \kappa) - n$	

Table 2.1. *Symmetric sigma points with parameters α and β*

We need to define the following parameters for this algorithm:

– *a scale factor* α that determines the spread of the initial sigma points around $\hat{\mathbf{x}}_{k|k}$. The parameter α is a positive real number, often chosen to be small (e.g. 10^{-1}). Values between 10^{-1} and 1 are typically found in the literature;

– *a scalar* κ that guarantees that the predicted covariance will be positive semidefinite. The inequality $n+\kappa > 0$ is used to parameterize the $2n+1$ sigma points. If $\kappa \in \mathbb{R}$ is chosen to be negative, the predicted covariance matrix is not necessarily positive semidefinite. Therefore, κ must be chosen to be positive to guarantee that the predicted covariance is positive semidefinite;

– *a real number* β chosen to capture information about moments of order greater than three. The value $\beta = 2$ is optimal when the distribution of $\mathbf{x}_k$ is assumed to be normally distributed.

1) Compute the evolution of the $p+1$ sigma points :

$$\mathcal{X}^{(i)}_{k+1|k} = \mathbf{f}(\boldsymbol{\mathcal{X}}^{(i)}_{k|k}, \mathbf{u}_k, k).$$

2) Compute the predicted mean value:

$$\hat{\mathbf{x}}_{k+1|k} = \sum_{i=0}^{p} \mathbf{W}^{(i)} \boldsymbol{\mathcal{X}}^{(i)}_{k+1|k}.$$

3) Compute the predicted covariance:

$$\mathbf{P}_{k+1|k} = \sum_{i=0}^{p} \mathbf{W}^{(i)} \left[\boldsymbol{\mathcal{X}}^{(i)}_{k+1|k} - \hat{\mathbf{x}}_{k+1|k}\right] \left[\boldsymbol{\mathcal{X}}^{(i)}_{k+1|k} - \hat{\mathbf{x}}_{k+1|k}\right]^T. \quad [2.17]$$

4) Compute the predicted measurement from these $p+1$ image points:

$$\mathbf{z}^{(i)}_{k+1|k} = \mathbf{h}(\boldsymbol{\mathcal{X}}^{(i)}_{k+1|k}, \mathbf{u}_k, k),$$

$$\hat{\mathbf{z}}_{k+1|k} = \sum_{i=0}^{p} \mathbf{W}^{(i)} \mathbf{z}^{(i)}_{k+1|k}.$$

Box 2.1. *Sigma-point prediction algorithm*

Next, the prediction steps for the mean and covariance are performed exactly as described in Box 2.1. The correction step of the scaled UKF differs very little, since it is based on the same equations [2.12] as linear Kalman filtering, namely:

$$\hat{\mathbf{x}}_{k+1|k+1} = \hat{\mathbf{x}}_{k+1|k} + \mathbf{K}_{k+1} \cdot (\mathbf{z}_{k+1} - \hat{\mathbf{z}}_{k+1|k}),$$

$$\begin{aligned}\mathbf{P}_{k+1|k+1} &= E[(\mathbf{x}_{k+1} - \hat{\mathbf{x}}_{k+1}) \cdot (\mathbf{x}_{k+1} - \hat{\mathbf{x}}_{k+1})^T \mid \mathbf{z}^{k+1}] \\ &= E[(\mathbf{x}_{k+1} - \hat{\mathbf{x}}_{k+1|k+1}) \cdot (\mathbf{x}_{k+1} - \hat{\mathbf{x}}_{k+1|k+1})^T] \\ &= \mathbf{P}_{k+1|k} - \mathbf{K}_{k+1} \cdot \mathbf{P}_{zz,k+1|k} \cdot \mathbf{K}_{k+1}^T,\end{aligned}$$

where

$$\mathbf{K}_{k+1} = \mathbf{P}_{xz,k+1|k} \cdot \mathbf{P}_{zz,k+1|k}^{-1}.$$

The only difference is how the covariances of the measurement are calculated. Here, they are derived from the sigma points:

$$\mathbf{P}_{zz,k+1|k} = V_{k+1} + \sum_{i=0}^{p} \mathbf{W}^{(i)} \left[\mathbf{z}_{k+1|k}^{(i)} - \hat{\mathbf{z}}_{k+1|k}\right] \left[\mathbf{z}_{k+1|k}^{(i)} - \hat{\mathbf{z}}_{k+1|k}\right]^T,$$

$$\mathbf{P}_{xz,k+1|k} = \sum_{i=0}^{p} \mathbf{W}^{(i)} \left[\mathcal{X}_{k+1|k}^{(i)} - \hat{\mathbf{x}}_{k+1|k}\right] \left[\mathbf{z}_{k+1|k}^{(i)} - \hat{\mathbf{z}}_{k+1|k}\right]^T.$$

Note that this algorithm does not compute any form of Jacobian or Hessian. This is one of the major advantages of this type of method.

2.6. Invariant observer theory

Much research has been conducted on nonlinear invariant observers over the past decade, most notably by P. Rouchon, S. Bonnabel and E. Salan [BON 08, BON 09a, MAR 10], who developed a constructive model based on a combination of differential geometry and group theory that can be used to create nonlinear filters for nonlinear system state estimation problems. The advantage of their method is that it preserves the physical properties and symmetries of the system. To design this kind of filter, we first apply various group definitions to find transformations of the dynamic model of the system with the objective of establishing a suitable mathematical formulation of the problem. Next, we implement a geometric framework by finding an invariant vector field and defining an invariant error. To fine-tune the dynamics of the filter, we calculate the gain values by linearizing the state error, which is now nonlinear. The linearization is no longer performed around an equilibrium point but around the identity element of a group. As we shall see, this enlarges the domain of convergence of the estimate and gives simplified expressions for the dynamics of the estimation error. The definitions given below offer a brief tour of the most important research on constructing invariant filters. Later in this chapter, we will apply this theory to a simple academic example (a non-holonomic car). We will return to this example in Chapter 4 to develop an inertial navigation system. For now, consider the following nonlinear state representation:

$$\begin{cases} \frac{d\mathbf{x}}{dt} = f(\mathbf{x}, \mathbf{u}) \\ \mathbf{y} = h(\mathbf{x}, \mathbf{u}) \end{cases} \quad [2.18]$$

In equation [2.18], the state vector $\mathbf{x}$ (respectively, the vector of inputs $\mathbf{u}$, the vector of outputs $\mathbf{y}$) belongs to some open set $\mathcal{X} \in \mathbb{R}^n$ (respectively, $\mathcal{U} \in \mathbb{R}^m, \mathcal{Y} \in \mathbb{R}^p, p \leq n$).

From this initial nonlinear representation, our first step is to define a group structure. This structure is then used to construct a series of transformations that fix the dynamics (of the system) whenever they act on the open sets defined above.

DEFINITION 2.1.– *(Transformation group). Let G be a Lie group and suppose that $\mathcal{T}$ is an open subset of $\mathbb{R}^q$. Given a transformation group $(\theta_g)_{g \in G}$ indexed by G with identity application θ_e (where e is the identity element of G) such that $(g, \xi) \in G \times \mathcal{T} \mapsto \phi_g(\xi) \in \mathcal{T}$, an action of this transformation group on $\mathcal{T}$ is defined as a differentiable chart with the following properties:*

– $\forall \xi \in \mathcal{T}, \theta_e(\xi) = \xi$;

– $\forall (g_1, g_2) \in G^2, \theta_{g_2} \circ \theta_{g_1}(\xi) = \theta_{g_2 g_1}(\xi)$.

DEFINITION 2.2.– *(Identification of the Lie group and the open set). Let $(\theta_g)_{g \in G}$ be a transformation group of full rank. In other words, $\forall g \in G$, $dim(Im(\theta_g)) = dim(G) = r$. If $dim(G) = r = dim(\mathcal{T}) = q$, then G and $\mathcal{T}$*

can be identified as follows by identifying the mapping θ indexed by G with the left- or right-multiplications:

$$\forall (g, \xi) \in (G \times \mathcal{T}), \; \theta_g(\xi) = \xi g^{-1} = R_{g^{-1}}(\xi) \quad or \quad \theta_g(\xi) = g\xi = L_g(\xi).$$

By definition 2.2, $(\theta_g)_{g \in G}$ is a diffeomorphism. In the following, we shall only consider full-rank group actions that satisfy $rg(\theta_g) = dim(G)$ and $dim(G) = dim(M)$. With these assumptions, definition 2.2 guarantees that we can identify the group G and the open set $\mathcal{T}$.

Analogously, when the state is represented by the expression in equation [2.18], where the state vector (respectively, input vector, output vector) belongs to an open set $\mathcal{X} \in \mathbb{R}^n$ (respectively, $\mathcal{U} \in \mathbb{R}^m$, $\mathcal{Y} \in \mathbb{R}^p, p \leq n$), we can consider the group transformation defined by *right* multiplication on $\mathcal{X} \times \mathcal{U} \times \mathcal{Y}$ as follows:

$$\phi : G \times (\mathcal{X} \times \mathcal{U} \times \mathcal{Y}) \rightarrow (\mathcal{X} \times \mathcal{U} \times \mathcal{Y})$$
$$(g, \mathbf{x}, \mathbf{u}, \mathbf{y}) \mapsto \phi_g(\mathbf{x}, \mathbf{u}, \mathbf{y}) = (\varphi_g(\mathbf{x}), \psi_g(\mathbf{u}), \rho_g(\mathbf{y})) = (\mathbf{X}, \mathbf{U}, \mathbf{Y}),$$

where $(\varphi_g, \psi_g, \rho_g)$ are three diffeomorphisms indexed by $g \in G$ for some Lie group G satisfying $dim(G) = dim(\mathcal{X}) = n$. The coordinates of these transformations must be defined in such a way that conjugating the state, input, and output variables by the corresponding action fixes all of the dynamics of the system, i.e. $\dot{\mathbf{X}} = f(\mathbf{X}, \mathbf{U})$ and $\mathbf{Y} = h(\mathbf{X}, \mathbf{U})$ satisfy:

DEFINITION 2.3.– *(G-invariance of the system and G-equivariance of the output). The dynamics of the system are said to be G-invariant if there exists $(\varphi_g, \psi_g)_{g \in G}$, $\forall (g, \mathbf{x}, \mathbf{u}) \in G \times (\mathcal{X} \times \mathcal{U})$:*

$$f(\varphi_g(\mathbf{x}), \psi_g(\mathbf{u})) = D\varphi_g(\mathbf{x}) \cdot f(\mathbf{x}, \mathbf{u}),$$

where $D\varphi_g(\mathbf{x})$ denotes the derivative mapping of φ_g at $\mathbf{x}$. The output of the system is said to be G-equivariant if $\exists (\rho_g)_{g \in G}$ $G \times \mathbf{y} \mapsto \mathbf{y}$, $\forall (g, \mathbf{x}, \mathbf{u}) \in G \times (\mathcal{X} \times \mathcal{U})$:

$$h(\varphi_g(\mathbf{x}), \psi_g(\mathbf{u})) = \rho_g(h(\mathbf{x}, \mathbf{u})).$$

This definition means that the evolution and observation equations must remain explicitly identical after applying the three transformations. Based on Cartan's method of moving frames (see Appendix A), a complete set of n-invariants in G can be constructed by considering the group action ψ_g.

DEFINITION 2.4.– *(Normalization equation). Given any two points* $x, c \in \mathcal{T}$, *there exists* g *such that either* $\theta_g(x) = L_g(x) = gx = c$ *or* $\theta_g(x) = R_{g^{-1}}(x) = xg^{-1} = c$. *This guarantees the existence of a moving frame* γ *at any point* $x \in \mathcal{T}$. *In particular, by choosing* $c = e$, *we deduce that* $\gamma(x) = x^{-1}$. *This procedure is known as "normalization".*

Remarks about invariants $I(\hat{\mathbf{x}}, \mathbf{u})$*:* Consider the composite transformation group defined by:

$$\begin{aligned}\phi : G \times (\mathcal{X} \times \mathcal{U} \times \mathcal{Y}) &\rightarrow (\mathcal{X} \times \mathcal{U} \times \mathcal{Y}) \\ (g, \mathbf{x}, \mathbf{u}, \mathbf{y}) &\mapsto \phi_g(\mathbf{x}, \mathbf{u}, \mathbf{y}) = (\varphi_g(\mathbf{x}), \psi_g(\mathbf{u}), \rho_g(\mathbf{y})) = (\mathbf{X}, \mathbf{U}, \mathbf{Y}),\end{aligned}$$

where $dim(G) = dim(\mathcal{X}) = n$ and φ_g has full rank. Then:

$$\begin{aligned}\phi : G \times (\mathcal{X} \times \mathcal{U} \times \mathcal{Y}) &\rightarrow (\mathcal{X} \times \mathcal{U} \times \mathcal{Y}) \in \mathbb{R}^{n+m+p} \\ (g, \mathbf{x}, \mathbf{u}, \mathbf{y}) &\mapsto \phi_g(\mathbf{x}, \mathbf{u}, \mathbf{y}) = (g\mathbf{x}, \psi_g(\mathbf{u}), \rho_g(\mathbf{y})) = (\mathbf{X}, \mathbf{U}, \mathbf{Y}).\end{aligned}$$

Now, consider a point with coordinates $c \in \mathrm{Im}(\varphi_g) \in \mathbb{R}^n$, and suppose that we wish to find the index $g \in G$ that satisfies:

$$\phi_g^{\mathbb{R}^n}(\mathbf{x}, \mathbf{u}, \mathbf{y}) = \varphi_g(\mathbf{x}) = g\mathbf{x} = c,$$

where $\phi_g^{\mathbb{R}^n}(\mathbf{x}, \mathbf{u}, \mathbf{y})$ denotes the restriction of ϕ_g to the components of the state $\mathbf{x}$ only. The objective when solving this problem, often described as solving the normalization equations, is to find the local transformation restricted to $\mathcal{X}$ that sends $\mathbf{x} \in \mathcal{X}$ to the transverse section $\mathcal{C}$ with orbits $(\mathbf{x}, \mathbf{u}, \mathbf{y})$ defined by $\mathcal{C} = \{c_1, c_2, ..., c_n\}$.

By the regularity properties of $\phi_g^{\mathbb{R}^n}$, and thus of φ_g (which is a diffeomorphism and so is invertible), we can apply the implicit function theorem to guarantee the existence of a function γ such that:

$$g \cdot \mathbf{x} - c = 0 \Leftrightarrow g = \gamma(\mathbf{x}).$$

This function γ gives the local definition of the moving frame at $\mathbf{x}$ that sends this point to the transverse section $\mathcal{C}$. In the special case where the Lie group G and the state space $\mathcal{X}$ are identified, replacing the index g by $\gamma(\mathbf{x})$

in the triple $(g\mathbf{x}, \psi_g(\mathbf{u}), \rho_g(\mathbf{y}))$ defines a complete set of $m + p$ functionally independent invariants such that:

$$I(\mathbf{x}, \mathbf{u}, \mathbf{y}) = \phi_{\gamma(\mathbf{x})}^{\mathbb{R}^{m+p}}(\mathbf{x}, \mathbf{u}, \mathbf{y}) = (\psi_{\gamma(\mathbf{x})}(\mathbf{u}), \rho_{\gamma(\mathbf{x})}(\mathbf{y})).$$

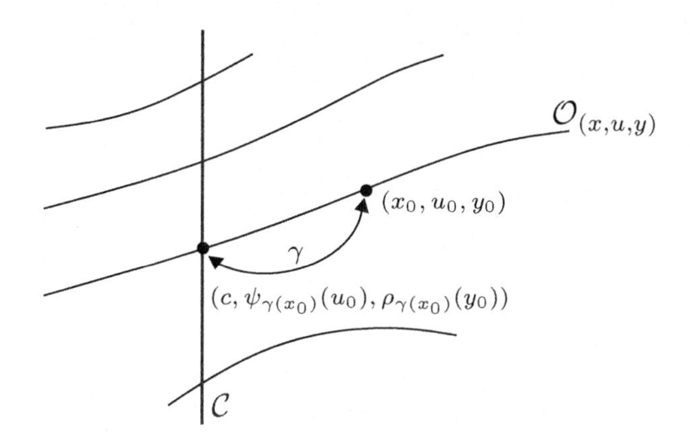

Figure 2.6. *Orbits of the transformation group*

Invariance means that $I(\phi_g(\mathbf{x}, \mathbf{u}, \mathbf{y})) = I(\mathbf{x}, \mathbf{u}, \mathbf{y}), \forall g, \mathbf{x}, \mathbf{u}, \mathbf{y}$. To show that I satisfies this property, we can start from:

$$\phi_g^{\mathbb{R}^n}(\mathbf{x}, \mathbf{u}, \mathbf{y}) = \varphi_{\gamma(\mathbf{x})}(\mathbf{x}) = c.$$

Then:

$$\forall g \in G,\ \varphi_{\gamma(\varphi_g(\mathbf{x}))}\big(\varphi_g(\mathbf{x})\big) = c = \varphi_{\gamma(\varphi_g(\mathbf{x}))g}(\mathbf{x}).$$

Since the moving frame is unique, we have that:

$$\forall g \in G,\ \forall \mathbf{x} \in \mathcal{X},\ \gamma\big(\varphi_g(\mathbf{x})\big)g = \gamma(\mathbf{x}).$$

Hence:

$$\begin{aligned} I\big(\phi_g(\mathbf{x}, \mathbf{u}, \mathbf{y})\big) &= \phi_{\gamma(\phi_g(\mathbf{x},\mathbf{u},\mathbf{y}))}^{\mathbb{R}^{m+p}}\big(\phi_g(\mathbf{x}, \mathbf{u}, \mathbf{y})\big) = \phi_{\gamma(\varphi_g(\mathbf{x}))g}^{\mathbb{R}^{m+p}}(\mathbf{x}, \mathbf{u}, \mathbf{y}) \\ &= \phi_{\gamma(\mathbf{x})}^{\mathbb{R}^{m+p}}(\mathbf{x}, \mathbf{u}, \mathbf{y}) = \big(\psi_{\gamma(\mathbf{x})}(\mathbf{u}), \rho_{\gamma(\mathbf{x})}(\mathbf{y})\big) = I(\mathbf{x}, \mathbf{u}, \mathbf{y}). \end{aligned}$$

In particular, when $c = e$ (the identity element), we find that:

$$\gamma(\mathbf{x}) = \mathbf{x}^{-1} \text{and } I(\mathbf{x}, \mathbf{u}, \mathbf{y}) = \left(\psi_{\mathbf{x}^{-1}}(\mathbf{u}), \rho_{\mathbf{x}^{-1}}(\mathbf{y})\right).$$

The $m + p$ fundamental invariants defined above will be used to construct invariant observers later.

DEFINITION 2.5.– *An invariant filter can be written as:*

$$\dot{\hat{\mathbf{x}}} = f(\hat{\mathbf{x}}, \mathbf{u}) + \sum_{i=1}^{n} \mathbf{K}_i(I(\hat{\mathbf{x}}, \mathbf{u}), E) \cdot w_i(\hat{\mathbf{x}}), \qquad [2.19]$$

where the gain matrix $\mathbf{K}_i$ only depends on the trajectory of the system over the complete set of known invariants $I(\hat{\mathbf{x}}, \mathbf{u}) = \psi_{\hat{\mathbf{x}}^{-1}}(\mathbf{u})$. The vector $w_i(\hat{\mathbf{x}}) := \left[D\varphi_{\gamma(\hat{\mathbf{x}})}(\hat{\mathbf{x}})\right]^{-1} \cdot \partial/\partial\mathbf{x}_i$ is an invariant vector that projects the set of invariant correction terms onto each component of the evolution equation $f(\hat{\mathbf{x}}, \mathbf{u})$, i.e. the tangent space, where $\partial/\partial\mathbf{x}_i$ is the ith vector field of the canonical basis of $\mathbb{R}^n$.

The convergence properties of equation [2.19] depend on the gain terms $\mathbf{K}_i$ and the definition of the estimation error. Instead of considering a conventional linear estimation error $\hat{\mathbf{x}} - \mathbf{x}$, invariant observer theory defines an invariant estimation error, written as $\eta(\mathbf{x}, \hat{\mathbf{x}}) = \mathbf{x}^{-1}\hat{\mathbf{x}}$.

DEFINITION 2.6.– *The value of $\hat{\mathbf{x}}$ convergences asymptotically to $\mathbf{x}$ if and only if the dynamics of the invariant state error governed by the following equation (stated in a very general form) are stable:*

$$\dot{\eta} = \Upsilon(\eta, I(\hat{\mathbf{x}}, \mathbf{u})), \qquad [2.20]$$

where Υ is a smooth function (i.e. a function with continuity class $\mathcal{C}^\infty$). Thus, η depends only on the trajectory of the system over the known invariants $I(\hat{\mathbf{x}}, \mathbf{u})$.

This method of construction can be applied systematically to design (pre)observers, also known as candidate observers, whose correction terms preserve the symmetries of the original system. However, the hypotheses on the regularity and properties of the group action on the state, namely $dim(\varphi_g) = dim(G)$, are an obstacle to this approach being truly general. As we shall see in Chapter 4, in the special case where the system dynamics are left-invariant on a Lie group with a right-compatible output, we can define a class of nonlinear observers that are locally convergent around any trajectory and whose global behavior is independent of the trajectory of the system.

2.6.1. *Illustrative example: non-holonomic car*

Consider a non-holomic car (see Figure 2.7) with two control inputs, namely the acceleration of the front wheels (assumed to be the driving wheels) and the rotation speed of the steering wheel. This system is parameterized as follows: δ is the angle between the front wheels and the axis of the car, θ is the angle between the axis of the car and the horizontal direction (Ox), and (x, y) are the coordinates of the midpoint of the rear axle of the car with respect to some reference frame. To model the hypothesis that the car is non-holonomic, the two driving wheels are assumed to spin at the same speed v. Thus, we can simply imagine two virtual wheels located at the midpoint of each axle. By the rule of composition of speeds, the dynamics of the car can be written as follows [LAU 01]:

$$\mathcal{M}\begin{cases} \begin{pmatrix} \dot{x} \\ \dot{y} \\ \dot{\theta} \end{pmatrix} = \begin{pmatrix} v\cos(\delta)\cos(\theta) \\ v\cos(\delta)\sin(\theta) \\ \dfrac{v\sin(\delta)}{L} \end{pmatrix} = \begin{pmatrix} u\cos(\theta) \\ u\sin(\theta) \\ uv \end{pmatrix} & \text{(evolution equation)} \\ h(x, y, \theta) = (x, y) & \text{(observation equation)} \end{cases} \quad [2.21]$$

where u is the velocity and v is a variable deduced from the turning angle δ and the distance L by the formula $v = \tan(\delta/L)$. First, note that Figure 2.7 shows that the system dynamics are independent of the origin and orientation of the reference frame. They are therefore invariant under the action of the group $SE(2)$ of planar rotations and translations (see Appendix A).

The state space $\mathcal{X} = \mathbb{R}^2 \times \mathcal{S}^1$ is topologically equivalent to $SE(2)$, as shown by Figure 2.8, where $T(t)$ denotes the trajectory of the car. This will ultimately mean that the dynamics fixed by equation [2.21] are invariant under the action of a transformation group that we can identify with left-multiplication in this example. Indeed, by identifying G with $\mathbb{R}^2 \times \mathcal{S}^1$, for any given vector $(x_0\ y_0\ \theta_0)^T \in G$, the mappings $\varphi_{(g_0)}(x)$ and $\psi_{(g_0)}(x)$ defined below give a left-multiplication action of G on itself that fixes the dynamics of the system. In the explicit model $\mathcal{M}$, the resulting image variables ($\varphi_{(g_0)}(x)$ and $\psi_{(g_0)}(x)$) allow us to conclude that the dynamics of the system are G-invariant.

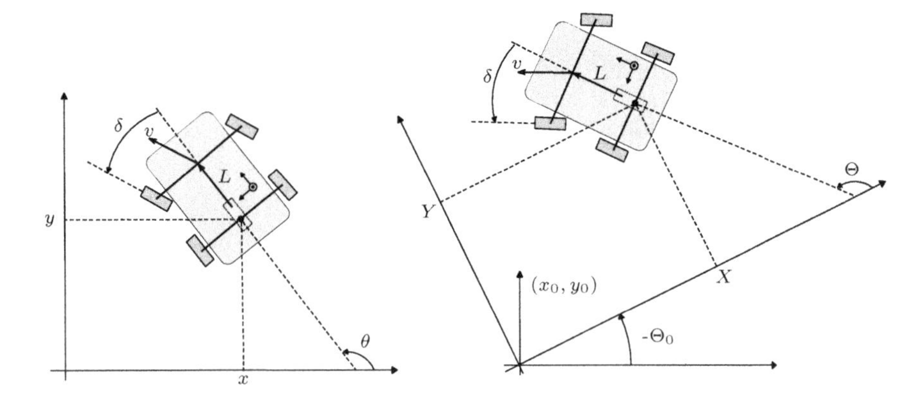

Figure 2.7. *The dynamics of the car are invariant under the action of the group $SE(2)$*

We have that $\forall g = (x_0\, y_0\, \theta_0) \in G$:

$$\varphi_g\left(x = (x\ y\ \theta)^T\right) = \begin{pmatrix} x_0 \\ y_0 \\ \theta_0 \end{pmatrix} \cdot \begin{pmatrix} x \\ y \\ \theta \end{pmatrix} = \begin{pmatrix} x \cos\theta_0 - y \sin\theta_0 + x_0 \\ x \sin\theta_0 + y \cos\theta_0 + y_0 \\ \theta + \theta_0 \end{pmatrix}$$

$$= \begin{pmatrix} X \\ Y \\ \Theta \end{pmatrix},$$

$$\psi_{(x_0,y_0,\theta_0)}(u, v) = \begin{pmatrix} u \\ v \end{pmatrix} = \begin{pmatrix} U \\ V \end{pmatrix}.$$

Thus, the dynamics of the system are indeed invariant in the sense of definition 2.3, since we can rewrite them as follows in terms of the new variables:

$$\begin{aligned}\dot{X} &= \overset{\cdot}{\overbrace{(x \cos\, \theta_0 - y \sin\, \theta_0 + x_0)}} = \dot{x} \cos \theta_0 - \dot{y} \sin \theta_0 \\ &= u \cdot (\cos\, \theta \cdot \cos\, \theta_0 - \sin\, \theta \cdot \sin\, \theta_0) \\ &= u \cdot \cos\,(\theta + \theta_0) \\ &= U \cos\, \Theta,\end{aligned}$$

$$\begin{aligned}\dot{Y} &= \overset{\cdot}{\overbrace{(x \sin\, \theta_0 - y \cos\, \theta_0 + y_0)}} = \dot{x} \sin \theta_0 - \dot{y} \cos \theta_0 \\ &= u \cdot (\cos\, \theta \cdot \sin\, \theta_0 - \sin\, \theta \cdot \cos\, \theta_0) \\ &= u \cdot \sin\,(\theta + \theta_0) \\ &= U \sin\, \Theta,\end{aligned}$$

$$\dot{\Theta} = \overset{\cdot}{\overbrace{(\theta + \theta_0)}} = \dot{\theta} = uv = UV.$$

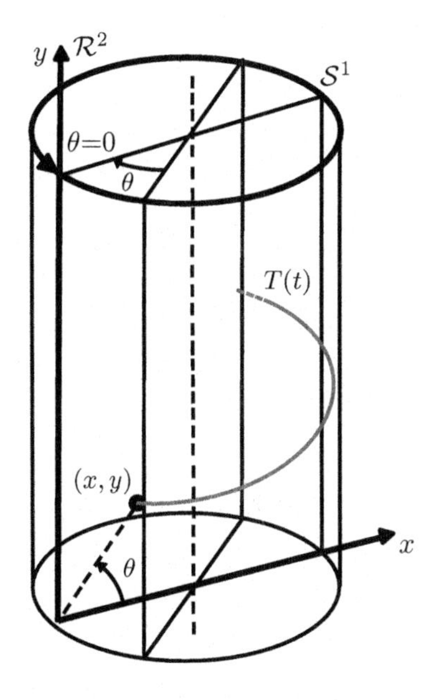

Figure 2.8. *The state space $\mathcal{X} = \mathbb{R}^2 \times \mathcal{S}^1$ is topologically equivalent to the group $SE(2)$*

Moreover, the observation equation is G-compatible in the sense of definition 2.3, since, for all $(x_0\ y_0\ \theta_0)^T \in G$, the image variable $\rho_{(g_0)}(x)$ satisfies

$$\rho_{(x_0,y_0,\theta_0)}(x,y) = \begin{pmatrix} x\cos\theta_0 - y\sin\theta_0 + x_0 \\ x\sin\theta_0 + y\cos\theta_0 + y_0 \end{pmatrix}.$$

This can be rewritten as follows in the new variables:

$$\begin{aligned} \rho_{(x_0,y_0,\theta_0)}(x,y) &= \begin{pmatrix} x\cos\theta_0 - y\sin\theta_0 + x_0 \\ x\sin\theta_0 + y\cos\theta_0 + y_0 \end{pmatrix} \\ &= h\Big(\varphi_g(x), \psi_g(u)\Big) = h(X,Y). \end{aligned}$$

Suppose now that we attach a trailer to the vehicle at the midpoint of the rear axle. From Figure 2.9 and equation [2.21], the new system has the following mathematical representation:

$$\mathcal{M}^+ \begin{cases} \begin{pmatrix} \dot{x} \\ \dot{y} \\ \dot{\theta} \\ \dot{\theta}_r \end{pmatrix} = \begin{pmatrix} v\cos(\delta)\cos(\theta) \\ v\cos(\delta)\cos(\theta) \\ \dfrac{v\sin(\delta)}{L} \\ \dfrac{v\cos(\delta)\sin(\theta-\theta_r)}{L_r} \end{pmatrix} = \begin{pmatrix} u\cos(\theta) \\ u\cos(\theta) \\ uw \\ u\sin(\theta-\theta_r) \end{pmatrix} \\ \\ h(x,y,\theta) = (x,y) \end{cases}$$

where the parameter L_r denotes the distance between the attachment point and the midpoint of the trailer's axle. It is clear that the variables $x_r, y_r, \delta_r, \dot{x}_r, \dot{y}_r, ...$ can be deduced analytically whenever both the state of the car and the angle θ_r are known.

This new state model $\mathcal{M}^+$ is left unchanged by the action of the group $SE(2)$. Indeed, for any $(x_0\ y_0\ \theta_0)^T \in G$, we can find a mapping $\varphi_{(x_0\ y_0\ \theta_0)}$ for which the dynamics of the model are G-invariant:

$$\varphi_g\left(x = (x\ y\ \theta\ \theta_r)^T\right) = \begin{pmatrix} x\cos\theta_0 - y\sin\theta_0 + x_0 \\ x\sin\theta_0 + y\cos\theta_0 + y_0 \\ \theta + \theta_0 \\ \theta_r + \theta_0 \end{pmatrix} = \begin{pmatrix} X \\ Y \\ \Theta \\ \Theta_r \end{pmatrix}.$$

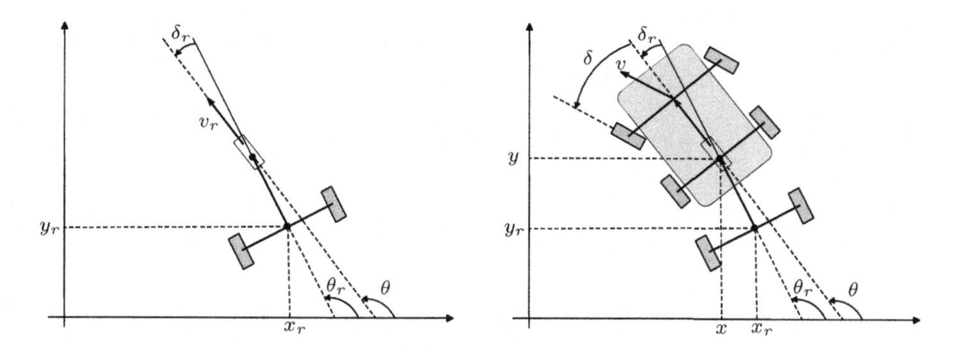

Figure 2.9. *A car towing a trailer*

Even though we introduced a state variable θ_r, the dynamics of the evolution equation are invariant in the sense of definition 2.3. The dynamics of θ_r can be rewritten as follows in the new variables:

$$\dot{\Theta}_r = \dot{\theta}_r = u \sin(\theta - \theta_r) = U \sin(\Theta - \Theta_r).$$

We can use this new transformation group to construct an invariant observer even when $dim(G) < dim(\mathcal{X})$. Below, we will apply the construction method introduced earlier to obtain an invariant observer. We will stop once we have found an invariant candidate observer (also known as a preobserver). In Chapter 4, we will discover why the convergence properties of equation [2.19] are useful – they will allow us to configure the correction gain K. This section aims to give a general understanding of each of the steps involved in constructing invariant preobservers in the context of an educational example. We will also analyze the properties of this example to demonstrate how the state estimation can sometimes be established analytically. The above example of a simple non-holonomic car is perfect for this.

2.6.1.1. *Constructing an invariant output error*

Consider again the model described by equation [2.21], which is G-invariant and G-equivariant. The transformation group defined earlier satisfies $dim(G) = dim(\mathcal{X}) = n = 3$, which allows us to identify G and $\mathcal{X}$. The group actions will also have full rank, i.e. $dim(\text{Im}(\varphi_g)) = dim(G) = 3$. After identifying G and $\mathbb{R}^2 \times S^1$ from the special case where

$g = x^{-1} = \big(-x\cos\theta - y\sin\theta,\ x\sin\theta - y\cos\theta,\ -\theta\big)^T$, we can deduce the following equality from the normalization equations (see definition 2.4):

$$\varphi_{x^{-1}}(x) = \begin{pmatrix} x\cos(-\theta) - y\sin(-\theta) - x\cos\theta - y\sin\theta \\ x\sin(-\theta) + y\cos(-\theta) + x\sin\theta - y\cos\theta \\ \theta - \theta \end{pmatrix} = \begin{pmatrix} 0 \\ 0 \\ 0 \end{pmatrix} = e, \tag{2.22}$$

where e is the identity element of the group G under addition. From the results of the group equation (equation [2.22]), we can then define the complete set of $m + p$ fundamental invariants using the parametrization $g = x^{-1} \in G$. The first m fundamental invariants for the dynamics of the system are obtained from $\psi_{x^{-1}}(u)$, and the remaining p fundamental invariants for the observation of this system are deduced from $\rho_{x^{-1}}(y)$. To construct the invariant output error, we no longer consider the usual linear output error $\hat{y} - y = h(\hat{x}, u) - y$, which does not preserve the symmetry of the system, but instead construct an invariant output error from the invariant output transformation $\rho_{(x_0,y_0,\theta_0)}(x, y)$ and the solution γ of the normalization equation (the moving frame) as follows:

$$\gamma(x) = \left(-x\cos\theta - y\sin\theta \ \ x\sin\theta - y\cos\theta \ \ -\theta\right)^T.$$

The following expression therefore defines an invariant output error:

$$\begin{aligned} E &= \rho_{\gamma(x_0,y_0,\theta_0)}(\hat{x}, \hat{y}) - \rho_{\gamma(x_0,y_0,\theta_0)}(x, y) \\ &= \begin{pmatrix} \cos\theta_0 & -\sin\theta_0 \\ \sin\theta_0 & \cos\theta_0 \end{pmatrix} \begin{pmatrix} \hat{x} \\ \hat{y} \end{pmatrix} + \begin{pmatrix} x_0 \\ y_0 \end{pmatrix} - \begin{pmatrix} \cos\theta_0 & -\sin\theta_0 \\ \sin\theta_0 & \cos\theta_0 \end{pmatrix} \begin{pmatrix} x \\ y \end{pmatrix} - \begin{pmatrix} x_0 \\ y_0 \end{pmatrix} \\ &= \begin{pmatrix} \cos\hat{\theta} & \sin\hat{\theta} \\ -\sin\hat{\theta} & \cos\hat{\theta} \end{pmatrix} \begin{pmatrix} \hat{x} - x \\ \hat{y} - y \end{pmatrix}. \end{aligned}$$

Thus, the invariant output error is consistent with the symmetries of the system and can be interpreted as a form of classical output error projected onto a Frenet coordinate system (multiplication by a rotation matrix). This result is the same as the result used to decouple the inputs/outputs of a trajectory tracking error. If we view $\hat{P}$ as the estimated position and P as the true position (see Figure 2.10), we obtain the following two outputs as errors:

$$e_{\parallel} = (\hat{P} - P) \cdot \vec{T} \quad \text{and} \quad e_{\perp} = (\hat{P} - P) \cdot \vec{N},$$

where:

$$\vec{T} = \begin{pmatrix} \cos\theta \\ \sin\theta \end{pmatrix} \quad \text{and} \quad \vec{N} = \begin{pmatrix} -\sin\theta \\ \cos\theta \end{pmatrix}.$$

This does indeed recover the same invariant output errors.

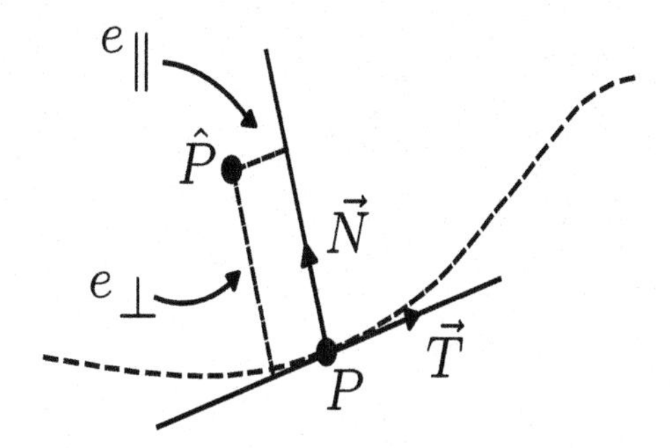

Figure 2.10. *Interpretation of the invariant error between the true position P and the estimate $\hat{P}$*

2.6.1.2. *Constructing an invariant coordinate system*

In three dimensions, we can find an invariant coordinate system (w_1, w_2, w_3) by taking the image of the standard basis of $\mathbb{R}^2 \times \mathcal{S}^1$ under the derivative mapping $DL_{(x,y,\theta)}$. In other words, we need to compute the derivative of the mapping $\varphi_g(x, y, \theta)$, which in this case is left-multiplication, in order to find the tangent space of the Lie algebra of $SE(2)$ (see Appendix A).

$$DL_{(x,y,\theta)} \begin{pmatrix} \frac{\partial X}{\partial x} & \frac{\partial X}{\partial y} & \frac{\partial X}{\partial \theta} \\ \frac{\partial Y}{\partial x} & \frac{\partial Y}{\partial y} & \frac{\partial Y}{\partial \theta} \\ \frac{\partial \theta}{\partial x} & \frac{\partial \theta}{\partial y} & \frac{\partial \theta}{\partial \theta} \end{pmatrix} = \begin{pmatrix} \cos\theta_0 & -\sin\theta_0 & 0 \\ \sin\theta_0 & \cos\theta_0 & 0 \\ 0 & 0 & 1 \end{pmatrix}. \qquad [2.23]$$

Thus, by inverting the Jacobian matrix from equation [2.23] at $\theta_0 = -\theta$ and identifying this matrix with the standard basis of $\mathbb{R}^3$, we obtain the following three invariant vector fields:

$$w_1 = \cos\theta\frac{\partial}{\partial x} + \sin\theta\frac{\partial}{\partial y}; \quad w_2 = -\sin\theta\frac{\partial}{\partial x} + \cos\theta\frac{\partial}{\partial y}; \quad w_3 = \frac{\partial}{\partial \theta}.$$

Note that this invariant coordinate system is equivalent to Frenet coordinates, which is not surprising, given the topology of the system trajectories acted upon by the Lie group $SE(2)$. In any Euclidean space (without curvature), the most natural observer for the purpose of estimation without any noise in the position $\hat{P}$ is of course the Luenberger observer, but in the case of a general manifold (such as a Riemmanian manifold), we need to account for the curvature (or geodesics) of the space.

In our example, to account for the geodesic ζ, the construction method replaces the classical error term by an invariant nonlinear term in order to conserve the symmetries, then projects the correction gain terms onto another coordinate system that is itself invariant (Frenet coordinates) to guarantee that the dynamics are symmetric under rotations (see Figure 2.11).

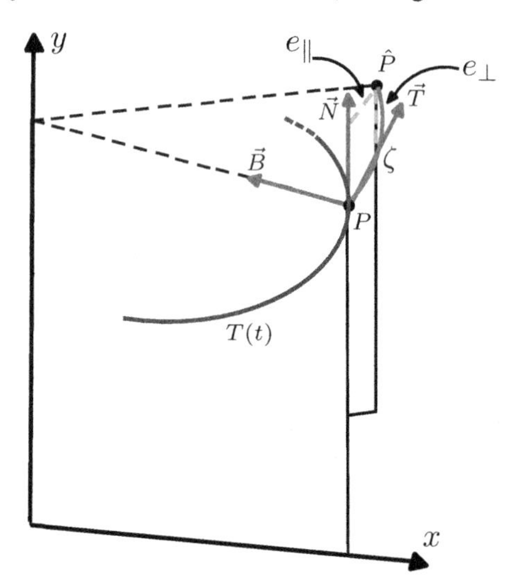

Figure 2.11. *Interpretation of the invariant coordinate system constructed on the group SE(2). For a color version of this figure, see www.iste.co.uk/condomines/kalman.zip*

2.6.1.3. *Invariant preobserver*

By definition 2.5, any invariant preobserver can therefore be written as:

$$\begin{pmatrix} \dot{\hat{x}} \\ \dot{\hat{y}} \\ \dot{\hat{\theta}} \end{pmatrix} = \begin{pmatrix} u\cos\hat{\theta} \\ u\sin\hat{\theta} \\ uv \end{pmatrix} + \begin{pmatrix} \cos\hat{\theta} & -\sin\hat{\theta} & 0 \\ \sin\hat{\theta} & \cos\hat{\theta} & 0 \\ 0 & 0 & 1 \end{pmatrix} \bar{\boldsymbol{K}} \begin{pmatrix} \cos\hat{\theta} & -\sin\hat{\theta} \\ \sin\hat{\theta} & \cos\hat{\theta} \end{pmatrix} \begin{pmatrix} \hat{x}-x \\ \hat{y}-y \end{pmatrix},$$

where $\bar{\boldsymbol{K}}$ is a 3×2 matrix of gain terms whose elements (in this particular case) depend only on the invariant output error E constructed earlier.

3

Inertial Navigation Models

3.1. Preliminary remarks: modeling mini-UAVs

When representing the dynamics of a controlled system such as the flight dynamics of an aerial vehicle mathematically, we often use the concept of state. By definition, the state of a system is the set of parameters whose values must be known at a given moment in time in order to predict the future evolution of the system. This idea is very natural for systems whose evolution over time can be described by differential equations. The solution of a system of differential equations of order $n \in \mathbb{N}^*$ depends on a set of n initial conditions. These initial values determine the subsequent states taken by the system over time. Thus, modeling dynamic systems by means of state representations, whether linear or nonlinear, tends to be more fruitful than other types of model (such as input/output black boxes), since it gives us access to a direct formulation of the underlying physics of the process. The role of an estimator is therefore to produce an accurate estimate of non-measured states (states that cannot be accessed directly by performing measurements) given knowledge of the inputs and an array of imperfect measurements. If our objective is to develop an accurate navigation system for a mini-UAV, the estimation process must be viewed as part of a more complex framework that needs to be optimized at every level. Figure 3.1 illustrates the position of Guidance, Navigation and Control (GNC) systems within this wider context, illustrating the complexity of the state estimation problem for autonomous aircraft. Each module is independent from the other, but each and every one of them significantly influences the flight parameters of the aircraft, as well as its ability to follow a reference trajectory. The estimation module inserts itself between the other modules as follows: the navigation system receives noisy measurements as inputs from the sensors, then merges these data using a dynamic model of the vehicle to compute a solution to the estimation problem. The estimated state can then be used either to generate a reference trajectory for the UAV to follow, from which a series of commands are deduced to control the aircraft (piloting and guidance), or to reconstruct flight parameters used in lower level control loops. Thus, the various measurements produced by the on-board sensors are processed by an advanced filtering algorithm that weights the imperfect measurements against the values predicted by a model. Estimation algorithms based on models of the dynamic system usually apply mechanical principles based on one of the following two advanced filtering techniques:

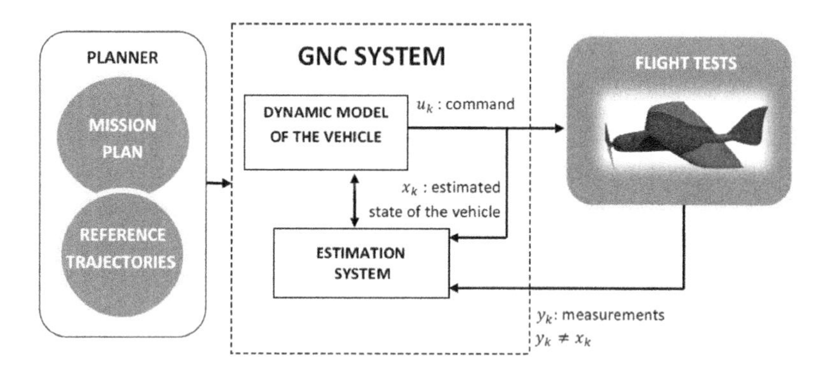

Figure 3.1. *Guidance, Navigation and Control (GNC) system for a UAV*

– *generic-model filtering* exploits our *a priori* knowledge of the classical laws of mechanics applied to a moving object without attempting to understand why the object is moving. Hence, a model of the net force acting on the aircraft is not required, because it is deduced from the measurements of a triaxial accelerometer. The advantage of this type of estimator is that it can be applied to any type of aerial vehicle without needing to adapt the algorithm, since the algorithm does not require a detailed model of the forces acting on the aircraft. The kinematics models used by these filters, often formulated in terms of quaternions for mini-drones given their agility in flight, allow the derivative of the state of the aircraft to be computed at any moment in time from a series of measured parameters and the current state. These models are therefore strongly dependent on the quality and continuous availability of the measurements provided by the on-board sensors. In the context of estimation, measurement inaccuracies are the greatest obstacle for this type of model. One obvious solution is to install better-performing sensors on the aircraft, subject to any constraints on the size of the aircraft and the development budget. Another solution is to re-estimate the inaccuracies online, which increases the complexity of the estimation problem by adding extra dimensions to the estimated state. This solution also encounters issues relating to the observability of the computed states, as we shall see later;

– *specific-model filtering* is based on physical models of the vehicle dynamics that include a description of the causes of motion (the external forces experienced by the aircraft). In the context of the flight dynamics of aerial vehicles, these forces include aerodynamic stresses, gravitational forces and propulsive stresses. The models are typically parametric and require an initial identification phase before operation. Hence, they heavily rely on our *a priori* knowledge. An understanding of the aerodynamics of the aircraft is usually gained by performing tests on a replica of the aircraft in a wind tunnel, or alternatively from Computational Fluid Dynamics (CFD) calculations (see Figure 3.2). The knowledge model needs to be as representative as possible of physical reality, so that we can use it both to simulate the behavior of the aircraft and to generate predictions as part of a model-based on-board estimation system. Establishing a satisfactory model is rarely straightforward, and ensuring that it is sufficiently representative of every aspect of the operational behavior of the true vehicle can be extremely challenging.

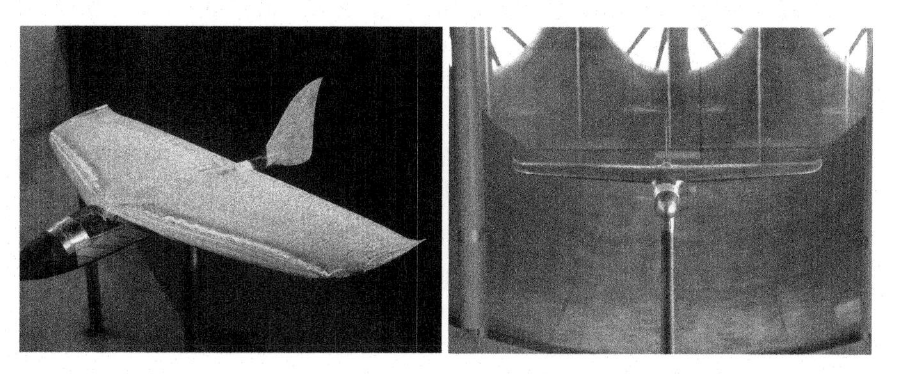

Figure 3.2. *Airflow over the wing of a UAV (ENAC/ISAE)*

Unlike purely kinematic models, any given specific model is only applicable to a single type of mini-UAV. Moreover, this approach only tends to work well with fixed-wing mini-UAVs (i.e. mini-airplanes) and has not or only somewhat been successfully applied to quad-rotors, whose unsteady aerodynamic behavior is complex to represent. In the context of estimation, and in contrast to filtering based on generic models, this highly physical and aircraft-specific category of model introduces new complications in the form of modeling errors, which are often tricky to handle. The nature of modeling errors is fundamentally very different from measurement errors (random noise or deterministic bias) and can cause the estimate to diverge. One common and straightforward solution is to express these errors as a series of additional state variables that are then re-estimated online, while ensuring that that the problem remains globally observable, similar to the second solution proposed above for handling measurement errors. Today, constant improvements in experimental technology and processing power have enabled us to develop increasingly reliable and accurate forecasting knowledge models. However, creating such a model demands significant investment (time, funding) that should be carefully considered before this modeling approach is adopted by a mini-UAV development project with "low-cost" constraints.

In this book, we study a series of different filters constructed from the same generic model. To design a complete high-performance avionic system (including both hardware and software) that we can use to control a mini-UAV, we need more than just high-performance control laws. We also require advanced filtering algorithms that can efficiently preprocess the data from the sensors and perhaps even reconstruct non-measured information or missing data. This is especially true in the context of mini-UAVs with low-cost instruments. More generally, any GNC system in a wider sense relies on having access to accurate estimates of the position, velocity, attitude angles and rotation speeds of the aircraft.

3.2. Derivation of the navigation model

Consider the following dynamic system (Σ), represented mathematically by an evolution model and an observation model:

$$(\Sigma) \begin{cases} \dot{\mathbf{X}} = f(\mathbf{X}, \mathbf{U}) \\ \mathbf{Z} = h(\mathbf{X}, \mathbf{U}) \end{cases}$$

In the state representation (Σ), $\mathbf{X}$ denotes the state vector, $\mathbf{Z}$ is the vector of outputs of the model and $\mathbf{U}$ is the vector of inputs. This general and abstract mathematical representation works equally well as a starting point for both generic and specific models. We can use it to construct state estimators that improve the navigation performance of aerial vehicles. Below, our objective is to construct a model that is as complete and sufficient as possible, and in particular capable of representing any navigational dynamics, such as the attitude, the velocity and the position of the aircraft, as accurately as possible. Although mini-UAVs usually have moving parts such as control surfaces that influence their pilotability and maneuverability, we will model them as a single body jointly acted upon by gravity and aerodynamic forces. By Newton's second law of motion:

$$m\dot{\vec{V_c}} = m\vec{g} + \vec{F} \quad [3.1]$$

$$\dot{\vec{\sigma_c}} = \vec{M_c} \quad [3.2]$$

In equations [3.1] and [3.2], m denotes the mass of the body, $\vec{V_c}$ is the velocity vector of the center of mass C, $\vec{g}$ is the acceleration vector due to gravity and $\vec{F}$ is the resultant of the aerodynamic forces. The parameter $\vec{\sigma_c}$ is the angular momentum of the aircraft about the point C, which fundamentally depends on the angular velocity vector $\vec{\Omega}$, and $\vec{M_c}$ denotes the moment of the external forces about C. The equations of the full model hold subject to the following two hypotheses:

HYPOTHESIS 3.1.– (Fixed-wing UAVs). The mass of the control surfaces is negligible relative to the total mass of the aircraft.

HYPOTHESIS 3.2.– (Rotating-wing UAVs). The rotating parts do not modify the position of the center of mass.

HYPOTHESIS 3.3.– (Flat Earth). The Earth is flat and does not rotate. We will always work in an inertial frame of reference.

The first and second hypotheses state that the navigation model is based on a rigid system with invariant mass. The third hypothesis states that the flight domain is negligibly small relative to the size of the Earth, and so the Earth's rotation does not influence the dynamics of the drone viewed from an inertial frame of reference. Thus, the complete model consists of equations [3.1] and [3.2], the kinematic relations between the position and the linear velocity, and the relations between the orientation and the angular velocity. The orientation of the mini-UAV is defined in terms of angles used to transition between so-called aircraft coordinates and ground coordinates. There are two options: the Euler angles φ, θ and ψ (yaw, pitch, roll) and quaternions. We will use quaternions below, since they avoid the singularity problems associated with the Euler angles and are somewhat robust against the various numerical errors that can potentially arise from repeatedly changing coordinates. In the most general setting that we need to consider, we must solve the problem of estimating the velocity, orientation and position of the mini-UAV. The equations of this model are as follows:

$$\dot{q} = \frac{1}{2} q * \omega \qquad [3.3]$$

$$\dot{V} = A + q * \frac{f}{m} * q^{-1} \qquad [3.4]$$

$$\dot{X} = V \qquad [3.5]$$

In equation [3.4], the external forces applied to the system are projected into ground coordinates (by the operation $q * \frac{f}{m} * q^{-1}$). This formulation of the equations therefore implicitly expresses the position, velocity and orientation of the aircraft in ground coordinates. We could instead have formulated each parameter in aircraft coordinates. We will use the following notation:

– q is the quaternion representing the orientation of the aircraft relative to the ground;

– V is the ground velocity vector of the aircraft, in ground coordinates;

– $A = (0\ 0\ g)^T$ is the acceleration vector due to gravity, in ground coordinates (North-East-Down [NED] convention);

– $\frac{f}{m}$ is the specific acceleration (caused by forces other than gravity) of the aircraft induced by the external forces acting upon the aircraft, expressed in aircraft coordinates.

As noted earlier, model-based estimation techniques follow one of two approaches (generic or specific) to represent the dynamic system. Each approach is associated with certain advantages and obstacles. In the context of the model presented above, we can be more specific.

– If the estimation algorithm is constructed from a specific model of the aircraft using equations [3.3]–[3.5], then an explicit model of the external forces f acting on the system is required.

– By contrast, estimation algorithms based on generic models apply equations that assume that the resultant force f is a known parameter obtained from a triaxial accelerometer. The accelerometer is attached to some point P on the aircraft and measures the parameter $\vec{a}$, the specific acceleration of the mini-UAV in aircraft coordinates, which satisfies $\vec{a} := \dot{\vec{V}_p} - \vec{g}$ at the point P. If the accelerometer is positioned at the center of mass C, then, by equation [3.1], it measures $\vec{a} = \dot{\vec{V}_p} - \vec{g} = \frac{1}{m}\vec{F}$; in other words, it measures $\frac{f}{m}$. However, if P is distinct from C, then the measurement is more difficult, since then $\vec{a} = \frac{1}{m}\vec{F} + \dot{\vec{\Omega}} \wedge \vec{CP} + \vec{\Omega} \wedge (\vec{\Omega} \wedge \vec{CP})$, and we need to include moments in the calculation.

Thus, in cases where we have inertial sensors (accelerometers and gyrometers giving the values of a_m and ω_m in aircraft coordinates), we can directly apply the following relations instead of equations [3.3]–[3.5]:

$$\mathcal{M}_s \begin{cases} \dot{q} = \frac{1}{2} q * \omega \\ \dot{V} = A + q * a * q^{-1} \\ \dot{X} = V \end{cases} \qquad [3.6]$$

In equation [3.6], the term V denotes the velocity vector at the point P, expressed in ground coordinates (replacing the term $\vec{V_P}$); this equation is in fact just the projection of $\vec{a} := \dot{\vec{V}_p} - \vec{g}$.

3.3. The problem of “true” inertial navigation

Armed with this description of a generic model, we can now solve the mathematical problem of estimating the state of the mini-UAV. First, we discuss whether it would be possible to develop an estimator based on the generic model (velocity/position/orientation) presented in section 3.2 simply by merging data from inertial sensors. The classical approach to constructing this type of estimator is to conduct an observability study (or more precisely a detectability study) of the state vector and vector of outputs (if it exists) to determine whether the state variables can be reconstructed from known system inputs and measurements. We shall see below that the system (Σ) has to satisfy certain conditions.

DEFINITION 3.1.– *(Observability of a dynamic system). The system* (Σ) *is said to be locally observable (about a point* $(\mathbf{x}, \mathbf{u}, \mathbf{y})$ *on some orbit) if the system state* $\mathbf{x}$ *can be expressed in terms of known signals* $\mathbf{u}, \mathbf{y}$ *and their derivatives of all orders. In other words:*

$$\mathbf{x} = \mathcal{M}(\mathbf{u}, \dot{\mathbf{u}}, \ddot{\mathbf{u}}, \cdots, \mathbf{y}, \dot{\mathbf{y}}, \ddot{\mathbf{y}}, \cdots).$$

If the system is not observable, then its observable and non-observable parts can be locally isolated and rewritten by performing a change of coordinates on the state as follows:

$$\dot{\mathbf{x}}_1 = f_1(\mathbf{x}_1, \mathbf{x}_2, \mathbf{u}) \qquad [3.7]$$

$$\dot{\mathbf{x}}_2 = f_2(\mathbf{x}_2, \mathbf{u}) \qquad [3.8]$$

$$\mathbf{y} = h(\mathbf{x}_2, \mathbf{u}), \qquad [3.9]$$

where equations [3.8] and [3.9] describe an observable subsystem. Assuming that an estimate of $\mathbf{x}_2$ is available from an observer, the only way to estimate the unobservable part $\mathbf{x}_1$ of the system is therefore to integrate equation [3.7] from an estimated initial condition $\hat{\mathbf{x}}_1(t = 0)$ such that $\dot{\hat{\mathbf{x}}}_1 = f_1(\hat{\mathbf{x}}_1, \hat{\mathbf{x}}_2, \mathbf{u})$. This will only give useful results if the dynamics f_1 are stable.

Now, if we consider the model $\mathcal{M}_s$, we see that the inertial measurements (a_m and ω_m) are inputs of the evolution model ($\dot{q}$ and $\dot{V}$). Since this model is not associated with an observation model, it is not observable in the sense of definition 3.1. The only way to estimate the state $(q^T \ V)^T$ is to rewrite the nonlinear equations of the modeled system[1] after replacing the true states by the estimates:

$$\mathcal{O}_s \begin{cases} \dot{\hat{q}} = \frac{1}{2}\hat{q} * \omega_m \\ \dot{\hat{V}} = A + \hat{q} * a_m * \hat{q}^{-1} \end{cases}$$

This gives an elementary observer without any correction terms, since $\mathcal{M}_s$ does not have an observation model. Thus, we need to analyze how the model behaves when there are errors, and in particular determine whether the system is detectable. To simplify the analysis, we perform the study on the system obtained by linearizing $\mathcal{M}_s$ about an equilibrium point $(\bar{q}^T\ \bar{\omega}_m^T\ \bar{V}^T\ \bar{a}_m^T)$. This linearization gives:

$$\begin{aligned} \delta\dot{q} &= \frac{1}{2}(\bar{q} * \delta\omega_m + \delta q * \bar{\omega}_m), \\ \delta\dot{V} &= \delta(q * a * q^{-1}) \\ &= \delta q * \bar{a}_m * \bar{q}^{-1} + \bar{q} * \bar{a}_m * \delta q^{-1} + \bar{q} * \delta a_m * \bar{q}^{-1} \\ &= \delta q * \bar{a}_m * \bar{q}^{-1} + \bar{q} * \bar{a}_m * \underbrace{(-\bar{q}^{-1} * \delta q * \bar{q}^{-1})}_{\delta q^{-1}} + \bar{q} * \delta a_m * \bar{q}^{-1}. \end{aligned}$$

Now, if we consider an equilibrium point near the state $(\bar{q}^T\ \bar{\omega}_m^T\ \bar{V}^T\ \bar{a}_m^T) = (\bar{q}^T\ 0\ 0\ \bar{a}_m^T)$, we find (note that, here, $\bar{a}_m = -\bar{q}^{-1} * A * \bar{q}$):

$$\begin{aligned} \delta\dot{q} &= \frac{1}{2}\bar{q} * \delta\omega_m, \\ \delta\dot{V} &= \delta q * (-\bar{q}^{-1} * A * \bar{q}) * \bar{q}^{-1} + \bar{q} * (-\bar{q}^{-1} * A * \bar{q}) * \underbrace{\delta q^{-1}}_{-\bar{q}^{-1} * \delta q * \bar{q}^{-1}} \\ &\quad + \bar{q} * \delta a_m * \bar{q}^{-1} \\ &= -\delta q * \bar{q}^{-1} * A + A * \delta q * \bar{q}^{-1} + \bar{q} * \delta a_m * \bar{q}^{-1} \\ &= -(\delta q * \bar{q}^{-1}) * A + A * (\delta q * \bar{q}^{-1}) + \bar{q} * \delta a_m * \bar{q}^{-1} \\ &= 2A \wedge (\delta_q * \bar{q}^{-1}) + \bar{q} * \delta a_m * \bar{q}^{-1}. \end{aligned}$$

1 The state associated with the position X is only considered when constructing the final observer, but this does not change the conclusions of our observability analysis for the full model.

In summary:

$$\begin{cases} \delta \dot{q} = \dfrac{1}{2}\bar{q} * \delta\omega_m \\ \delta \dot{V} = 2A \wedge (\delta_q * \bar{q}^{-1}) + \bar{q} * \delta a_m * \bar{q}^{-1} \end{cases}$$

We can now establish the linear system of estimation errors by defining $(\delta\eta, \delta\nu) = (\delta\hat{q} - \delta q, \delta\hat{V} - \delta V)$ and dividing into three subsystems (horizontal, lateral, vertical) to study the detectability of the estimator more precisely. The corresponding linearized observer has the equations:

$$\begin{aligned} \delta\dot{\eta} &= \delta\dot{\hat{q}} - \delta\dot{q} = \frac{1}{2}\bar{q} * \delta\omega_m - \frac{1}{2}\bar{q} * \delta\omega_m = 0, \\ \delta\dot{\nu} &= 2A \wedge (\delta\hat{q} * \bar{q}^{-1}) + \bar{q} * \delta a_m * \bar{q}^{-1} - 2A \wedge (\delta q * \bar{q}^{-1}) - \bar{q} * \delta a_m * \bar{q}^{-1} \\ &= 2A \wedge ((\delta\hat{q} - \delta q) * \bar{q}^{-1}) \\ &= 2A \wedge (\delta\eta * \bar{q}^{-1}). \end{aligned}$$

Requiring the quaternions to have norm one then yields the following system of differential equations:

$$\begin{cases} \delta\dot{\eta} = 0 \\ \begin{pmatrix} \delta\dot{\nu}_1 \\ \delta\dot{\nu}_2 \\ \delta\dot{\nu}_3 \end{pmatrix} = \begin{pmatrix} -2g\delta q_2 \\ 2g\delta q_1 \\ 0 \end{pmatrix} \end{cases} \Leftrightarrow \begin{cases} \delta\eta(t) = constant = \delta\eta_0 = \begin{pmatrix} \delta\hat{q}_0 - \delta q_0 \\ \delta\hat{q}_1 - \delta q_1 \\ \delta\hat{q}_2 - \delta q_2 \\ \delta\hat{q}_3 - \delta q_3 \end{pmatrix} \\ \begin{pmatrix} \delta\nu_1 \\ \delta\nu_2 \\ \delta\nu_3 \end{pmatrix} = \begin{pmatrix} -2g\delta q_2^0 t + \delta\nu_1^0 \\ 2g\delta q_1^0 t + \delta\nu_2^0 \\ \delta\nu_3^0 \end{pmatrix} \end{cases} \quad [3.10]$$

We can immediately see from equation [3.10] that the horizontal and lateral components of the errors in the velocity diverge linearly over time, whereas the error in the vertical velocity remains constant. The errors in the position, on the other hand, diverge parabolically, except for the error in the vertical position, which diverges linearly. The errors would be significantly worse if the model equations included any imperfections in the inertial measurements used in the calculation ($a_m + \omega_m$). In fact, we would have parabolic divergence in the horizontal and lateral components of the velocity error and linear divergence in the vertical velocity error. The position errors would also diverge at a rate of t^3 or t^2. Clearly, our conclusions would be completely different if we rewrote the equations of the system (equations [3.1] and [3.2]) to account for the curvature of the Earth[2]. In this case, the estimator would give bounded estimation errors for the horizontal and lateral velocities, which would oscillate with a period of around 84 minutes; this is known as the Schuler effect, and is used to define accuracy specifications for sensors (gyrometers and accelerometers). As a result, "low-cost" sensors are incompatible with this particular type of estimator, since the errors that they generate are simply too large. If the subsystem in the horizontal plane produces bounded errors, then the heading diverges linearly, and the altitude diverges exponentially. The model of the gravitational force (g is inversely proportional to z) causes the vertical velocity to diverge exponentially; recalibration with an altitude sensor (a baroaltimeter) is therefore necessarily required in order for the estimator to function properly.

This already gives us a perfectly valid inertial navigation system. However, we can only take advantage of the Schuler effect if our sensors are of sufficiently high performance and stable for the inaccuracy to remain small over the oscillation period of the effect (which is around 84 minutes). Unfortunately, the MEMS sensors used in mini-UAVs are far from meeting these requirements. As a result, the flat Earth hypothesis is sufficient to develop an estimation scheme that merges measurements from multiple sources and recalibrates the prediction model with suitable correction terms.

3.4. Modeling and identifying the imperfections of inertial sensors

As we noted earlier in section 1.3, inertial sensors based on MEMS technology are significantly better than mechanical and optical systems in terms of weight, size and cost. In terms of accuracy, however, they are not quite so excellent. Inaccuracies in MEMS sensors can arises from either deterministic errors or random errors. When MEMS technology was first

2 In practice, the Earth is modeled as ellipsoid to improve the accuracy. Thus, Hypothesis 3 is dropped by introducing a latitude coordinate to identify the position of a point on the surface of the Earth and writing the gravity field as a function of the altitude.

introduced, deterministic errors were more common and much more of a concern than they are today. Originally, to implement a triaxial gyrometer, three independent circuits mounted orthogonally in the navigation system were required. This design created significant balance issues between the axes and limited the maximum accuracy of the sensors. Today, the vast majority of MEMS gyrometers are manufactured on a single electronic chip, allowing deterministic errors, for example caused by alignment defects, to be calibrated once and for all by the manufacturers. Even so, these errors still tend to be underestimated, as argued in [DEA 13]. But inertial sensors are also affected by various types of random error. Typically, the inertial sensors measure the acceleration and the angular velocity of an object, then the measurements are integrated to calculate the position and the velocity of the object. However, these measurements capture more than just the motion of the object; they are influenced by various random phenomena, such as measurement drifts over time caused by temperature variations, or outliers induced by electronic noise, as well as other errors related to the aging of the sensor.

Hence, establishing an accurate model to account for random errors is a key objective when designing a high-performance estimation scheme. There are several ways to identify the various characteristic errors of MEMS sensors to various degrees of accuracy. The Allan variance technique presented below is the most widely used approach to modeling and identifying the errors of these sensors. It was originally developed to study the stability of precision oscillators, but can be applied systematically to any inertial system, and its results, computed offline, are straightforward to interpret. However, to achieve sufficient accuracy, it requires a large body of data.

3.4.1. *Allan variance*

The Allan variance method is widely used in the field of metrology. It computes the variance in the output signal of a sensor over time from a series of noisy measurements recorded by this sensor. The data are viewed as a vector of N datapoints sampled at intervals of τ_0, and divided into multiple groups, known as clusters. Each cluster represents a window of τ seconds consisting of exactly n samples $(n < N/2)$ (see Figure 3.3).

The method then computes the variance of the signal at multiple timescales. First, the variance is computed for the elementary timescale τ_0, then for every multiple of this timescale. The computation is performed with the clusters described above; the samples in each cluster are averaged before being used in the variance calculation. The results can then be presented as a log-log diagram plotting the variation of the standard deviation normalized by the average value of the signal as a function of the cluster size, revealing various characteristic imperfections in the signal (noise, periodic noise, drifts, etc.) as a function of the timescale τ of the variance calculation (see Figure 3.4). For gyroscopic or accelerometric inertial sensors, there are five major categories of noise: Quantization Noise (QN), Angular Random Walk (ARW), Bias Instability (BI), Rate Random Walk (RRW) and Rate Ramp (RR).

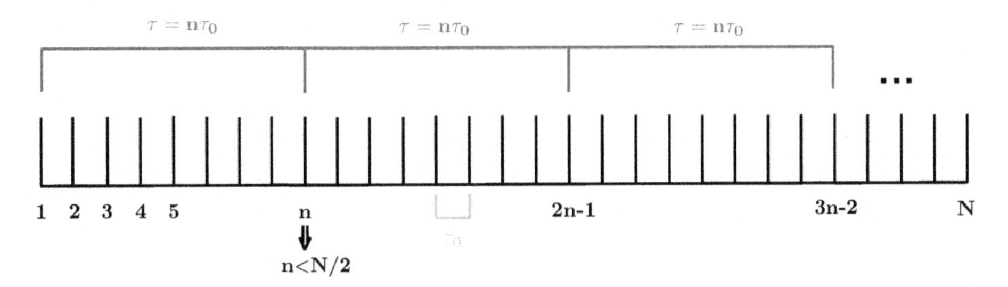

Figure 3.3. *Data structure for the Allan variance method. For a color version of this figure, see www.iste.co.uk/condomines/kalman.zip*

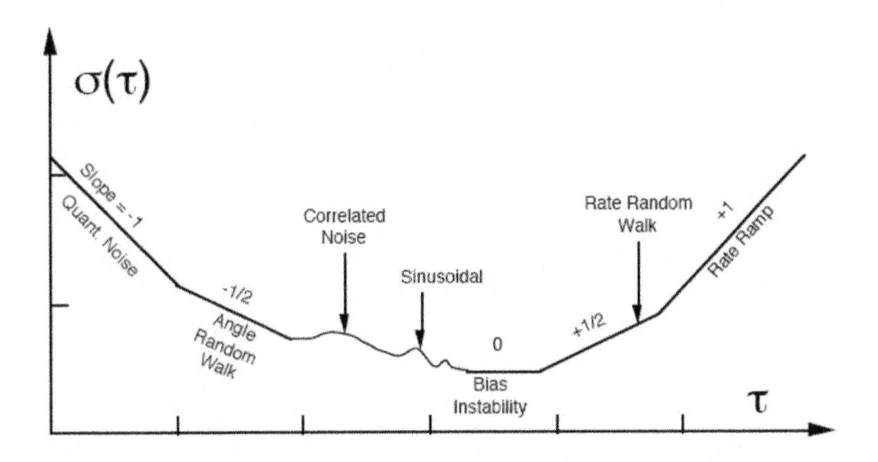

Figure 3.4. *Typical Allan variance curve for a MEMS inertial sensor*

To compute the characteristic variance of each type of noise in the signal, we first compute the mean $\Omega(t)$ of each of the M clusters of r samples as follows:

$$\forall k \in [\![1, M]\!], \bar{\Omega}_k(t) = \frac{1}{\tau} \int_{t_k}^{t_k+\tau} \Omega_k(t) dt. \qquad [3.11]$$

In equation [3.11], the term $\bar{\Omega}_k(t)$ denotes the mean of the signal for the k-th cluster. Unlike the classical variance, the Allan variance is defined as the average of the variance between consecutive terms of a sequence of samples $\{s_i\}_{0 \leq i \leq N}$ as follows:

$$\sigma^2_{Allan} = \frac{1}{2N} \sum_{k=0}^{N-1} (s_{i+1} - s_i)^2. \qquad [3.12]$$

The sequence of averages is classically defined as:

$$\eta_{k+1,k} = \bar{\Omega}_{k+1}(\tau) - \bar{\Omega}_k(\tau).$$

The Allan variance is deduced by substituting this sequence into equation [3.12] instead of the distance term $s_{i+1} - s_i$, i.e.:

$$\sigma^2(\tau) = \frac{1}{2(M-1)} \sum_{k=1}^{N-1} \eta^2_{k+1,k}.$$

The Allan variance can be shown to satisfy the following relation with the power spectral density of the signal:

$$\sigma^2(\tau) = 4 \int_0^{+\infty} S(f) \frac{\sin^4(\pi f t)}{(\pi f t)^2} df. \qquad [3.13]$$

The Allan variance can be approximated locally on a segment of timescales from the ARW term, which captures any high-frequency measurement noise in the sensor output, and the RRW term, which represents the long-term fluctuation in the measurement bias offset. We can therefore use these two terms to model the imperfections of the accelerometer and the gyrometer. The output of an inertial sensor (accelerometer or gyrometer) is influenced by multiple types of noise simultaneously; we shall assume that each type is statistically independent of the others. Thus, the Allan variance $\sigma^2(\tau)$ computed above can be explicitly written as:

$$\sigma^2(\tau) = \sigma^2_{QN}(\tau) + \sigma^2_{ARW}(\tau) + \sigma^2_{BI}(\tau) + \sigma^2_{RRW}(\tau) + \sigma^2_{RR}(\tau),$$

namely as a sum of five contributions from each type of noise. Next, using equation [3.13], we can calculate each term $\sigma^2_{\mathbf{x}}(\tau)$ in the Allan variance of the signal analytically by defining a power spectral density $S(f)$ for each type of noise. This allows us to deduce [ALL 66] [ALL 87]:

– $\sigma_{QN}(\tau) = Q\frac{\sqrt{3}}{\tau}$, where Q is the amplitude of the quantization noise, and can be read from the graph of $\sigma^2(\tau)$ at $\tau = 3^{1/2}$. The slope of this noise is -1 on log scales;

– $\sigma_{ARW}(\tau) = \frac{Q}{\sqrt{\tau}}$, where Q is the square root of the spectral density of the white noise characterizing the spectrum of this type of error. The ARW can be identified on the graph of $\sigma^2(\tau)$ by its envelope with slope $-1/2$;

– $\sigma_{BI}(\tau) \cong 0.6645 \cdot B$, where B is the instability coefficient of the bias in the expression of the spectral density associated with this noise[3]. It satisfies:

$$S(f) = \begin{cases} B^2/2\pi f \text{ for } f \leq f_0 \\ 0 \text{ otherwise} \end{cases}$$

This type of noise therefore adds a constant contribution to the Allan variance. Asymptotically, it takes the form of a constant region on the graph of $\sigma^2(\tau)$;

– $\sigma_{RRW}(\tau) = K\sqrt{\frac{\tau}{3}}$, where K is the square root of the spectral density of the noise at $\omega = 1 rad \cdot s^{-1}$. Its value can be read from the graph of $\sigma^2(\tau)$ at $\tau = 3$. This source of perturbation makes $\sigma^2(\tau)$ asymptotically tangent to a line with slope $+1/2$;

3 The power spectral density has bounded support and a spectrum shaped like $1/f$.

$-\sigma_{RR}(\tau) = R\frac{\tau}{\sqrt{2}}$, where R is the slope of the drift (known as the following error). The value of R can be read from the graph of $\sigma^2(\tau)$ at $\tau = \sqrt{2}$. Asymptotically, the Allan variance is tangent to a line with slope $+1$ on a log scale when $\tau >> 1$.

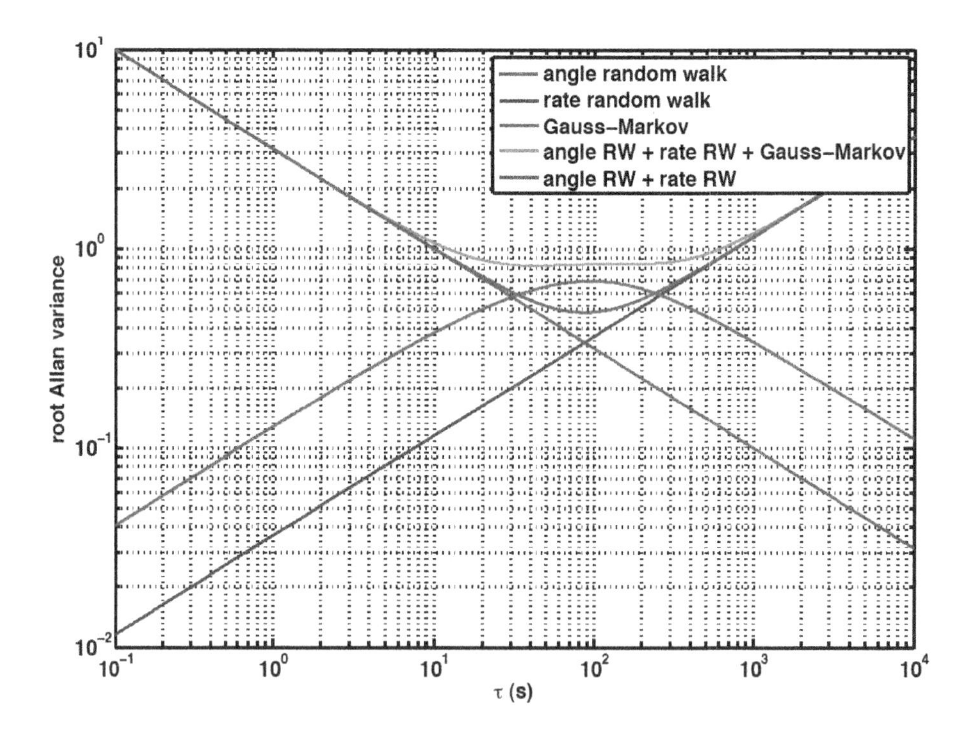

Figure 3.5. *Asymptotes of the Allan variance in the presence of different types of noise. For a color version of this figure, see www.iste.co.uk/condomines/kalman.zip*

3.4.2. *Modeling characteristic noise*

Building on the Allan variance method presented above, our next objective is to establish a model for the imperfections of an inertial sensor in order to represent their imperfections as accurately as possible. A general model can be written as follows [WEN 03, BRO 13]:

$$\zeta_m = (1 + S_f)\zeta_t + \nu_\zeta + b(t), \quad [3.14]$$

where ζ_m is the measurement produced by the sensor and ζ_t is the "true" measurement at time t. The term S_f is a scale factor. Here, we shall assume that the sensors were calibrated before being used, which means that the scale factor is zero, i.e. $S_f = 0$. The term $b(t)$ represents any measurement bias that varies over time, and ν_ζ is high-frequency noise, assumed to behave like white noise with a power spectral density R. The bias $b(t)$ could in turn be further broken down into the sum of two contributions: a constant component determined offline by calculating the mean of the sensor measurements over an extended period of time, and another component that varies slowly over time.

We can thus write:

$$b(t) = b_0 + b_1(t),$$

where $b_1(t)$ follows a Gauss–Markov process of the form:

$$\dot{b}_1(t) = \frac{-b_1(t)}{\tau} + \nu_{\zeta b_1}.$$

In the above expression, τ is a time constant specifying the dynamics of the filtered random walk, and $\nu_{\zeta b_1}$ is a Gaussian white noise generator with spectral density Q, which satisfies $E\{\nu_{\zeta b_1}(t_1)\nu_{\zeta b_1}(t_2)\} = Q\delta(t_1 - t_2)$. The three parameters of this model can be approximated by the Allan variance as shown in Figure 3.5. For our purposes, we shall work in the special case where $\tau = \infty$, which simplifies the previous model but still allows us to reasonably approximate the characteristics of the model describing the imperfections in the inertial sensors.

3.5. Inertial navigation on low budgets: AHRS

Suppose that the only proprioceptive sensors to which we have access are MEMS sensors that do not provide any information about the velocity or the position of the aircraft, as is indeed the case for many mini-UAVs. Ultimately, we must accept that we cannot estimate these quantities, even though they would be extremely useful for navigation. Instead, in order to estimate at the very least the attitude of the aircraft from the equations of the estimator $\mathcal{O}_s$, we need an additional hypothesis on its acceleration, as well as an exteroceptive sensor to bound the error in the estimated heading. To do this, our nonlinear state estimator will use three triaxial sensors that produce a total of nine scalar measurements:

– three measurements from three magnetometers (exteroceptive sensors) that give a local measurement of the Earth's magnetic field B, assumed to be known and constant[4] and expressed in aircraft coordinates as $y_B = q^{-1} * B * q$, where $B = [B_x \; B_y \; B_z]^T$ can be viewed as a model output;

– measurements of the angular velocity $\omega_m \in \mathbb{R}^3$ of the aircraft along three axes from three gyrometers. The angular velocity is written in the form $\omega_m = [\omega_{mx} \; \omega_{my} \; \omega_{mz}]^T$;

– finally, three measurements of the specific acceleration $a_m \in \mathbb{R}^3$ of the mini-UAV, written in the form $a_m = [a_{mx} \; a_{my} \; a_{mz}]^T$, delivered by three accelerometers.

We now make the following additional hypothesis to construct the estimator:

HYPOTHESIS 3.4.– (Acceleration of the mini-UAV). For the remainder of this section, we shall assume that the acceleration of the mini-UAV satisfies $||\dot{V}|| << ||A||$. Hence, from the kinematic relations presented earlier, we have that:

$$a = q^{-1} * (\dot{V} - A) * q \approx -q^{-1} * A * q.$$

If this hypothesis holds and the vector $A = [0 \; 0 \; g]^T$, which is assumed to be constant, is perfectly known, then the relation derived from this hypothesis above identifies a as an additional model output that can be compared against the measurement a_m in aircraft coordinates. Thus, the problem is reduced to estimating the orientation q based on the following model $\mathcal{M}_{ahrs}$:

$$\mathcal{M}_{ahrs} \begin{cases} \dot{q} = \frac{1}{2} q * \omega_m \\ y = \begin{pmatrix} y_A \\ y_B \end{pmatrix} = \begin{pmatrix} q^{-1} * A * q \\ q^{-1} * B * q \end{pmatrix} \end{cases} \qquad [3.15]$$

4 This value can be deduced from a model of the Earth's magnetic field; for example, the magnetic field B is equal to [0.5156 0.0570 0.8549] at the local flight coordinates 43.617-43°-37′ (N) and 1.450-1°-27′ (E) for the ENAC mini-UAVs used in this book.

A system that estimates the orientation from a combination of low-accuracy sensors and a magnetometer is often described as an Attitude and Heading Reference Systems (AHRS) in the field of aeronautics.

REMARK 3.1.– If the measurements of the inertial sensors are perfect (free from noise or bias), the state q can be deduced immediately. From the two output relations of equation [3.15], we can find the action of the rotation matrix associated with q without needing to apply the differential equation describing the dynamics $\dot{q}$. Hence, the attitude of the aircraft is fully determined.

3.5.1. *Observability of the underlying model of the AHRS*

In our case, we only have access to limited performance and thus imperfect sensors for our AHRS. In general, the primary sources of inherent error in navigation systems can be classified into the following three major categories:

1) low-level imperfections (noise, bias, drift, etc.) in the measuring instruments. We discussed the primary characteristics of this type of error in Chapter 1;

2) calibration-related errors, e.g. due to magnetic perturbations from the environment;

3) problems related to the gravitational model used by the navigation model.

Here, we are primarily interested in any types of error caused by intrinsic flaws in the sensors that can be modeled and integrated into the global model of the AHRS to improve the quality of the estimate. The three measuring instruments needed to develop the AHRS typically have two types of significant defect: instability in the measurement bias (expressed in degree/h for gyrometers and μg for accelerometers), and instability in the scale factor (see equation [3.14]). Both of these defects are approximately constant over the course of an experiment. However, it seems more sensible to account for the fact that they can vary slightly over time. It would be ideal to be able to estimate these errors online in parallel to the other states of the aircraft.

To determine the number of imperfections that we can expect to be able to model without rendering the global model unobservable, we can perform a first-order observability analysis of the equations of $\mathcal{M}_{ahrs}$ by linearizing this model around an arbitrary equilibrium point $(\bar{q}, \bar{\omega})$. This gives the following linearized model:

$$\begin{cases} \delta\dot{q} &= \frac{1}{2}(\bar{q} * \delta\omega + \delta q * \bar{\omega}) \\ \begin{pmatrix} \delta y_A \\ \delta y_B \end{pmatrix} &= \begin{pmatrix} \delta q * A * \bar{q}^{-1} + \bar{q} * A * (-\bar{q}^{-1} * \delta q * \bar{q}^{-1}) \\ \delta q * B * \bar{q}^{-1} + \bar{q} * B * (-\bar{q}^{-1} * \delta q * \bar{q}^{-1}) \end{pmatrix} \end{cases}$$

If we choose the value $(1, 0)$ for $(\bar{q}, \bar{\omega})$, we obtain the following linearized system, after assuming that the quaternion q has norm one, and noting that the equation $B_y \simeq 0$ holds in practice (the lateral deviation in the Earth's magnetic field is zero – the phenomenon of magnetic declination is neglected, i.e. we assume that geometric North and magnetic North coincide).

$$\begin{cases} \delta\dot{q} = \frac{1}{2}\delta\omega \\ \begin{pmatrix} \delta y_A \\ \delta y_B \end{pmatrix} = \begin{pmatrix} \delta q * A + A * \delta q \\ \delta q * B + B * \delta q \end{pmatrix} = \begin{pmatrix} 2A \times \delta q \\ 2B \times \delta q \end{pmatrix} = \begin{pmatrix} -2g\delta q_2 \\ 2g\delta q_1 \\ 0 \\ -2B_z\delta q_2 \\ 2(B_z\delta q_1 - B_x\delta q_3) \\ 2B_x\delta q_2 \end{pmatrix} \end{cases} \quad [3.16]$$

The model thus obtained has 4 states $[q_0\ q_1\ q_2\ q_3]^T$ and six outputs; three outputs for y_A and three outputs for y_B. In practice, we assume that the state δq_0 is related to the other states by the formula $\delta q_0 = \sqrt{1 - \delta q_1^2 - \delta q_2^2 - \delta q_3^2}$. Hence, the model will be observable by definition whenever $\delta q_1, \delta q_2$, and δq_3 can be expressed as a function of u, y, and their derivatives of all orders. This naturally holds in the first-order case by equation [3.16]. Therefore, from the three system inputs in the vector $\delta\omega_m$, which is assumed to be known, and from the six measurements associated with the vectors y_{Am} and y_{Bm} (i.e. nine values in total), we can guarantee that the system will remain observable if and only if a maximum of $9 - 3 = 6$ imperfections are introduced into the state and estimated. These imperfections will be measurement errors that are constant or which vary slowly over time; their dynamics can be modeled by random walks. The best choice for each of the six degrees of freedom varies depending on the specific application. In our case, we systematically sought to model three additive measurement biases in the gyrometers, i.e. biases that satisfy $\omega = \omega_m - \omega_b$, as will become clear later. But for now, we shall choose to estimate six measurement biases, one for each of the errors in the six components of the full output vector $(y_A^T \; y_B^T)^T$. The relations of equation [3.16] allow us to draw conclusions about the observability of each of these errors. Indeed, we cannot estimate all of the nine unknowns of the problem from just six relations, especially since the structure of the equations makes some of these biases impossible to observe. We shall therefore distinguish between two cases when establishing the observability of the model: imperfections in the vector of accelerometric measurements (y_A) and imperfections in the vector of magnetometric measurements (y_B).

CASE 1: maximize the number of imperfections modeled in the magnetometers

In this case, we can estimate two constant biases in y_B and one constant bias in y_A. In the special case of additive measurement biases satisfying $y_A = a_m = -a - a_b$ and $y_B = b_m = b - b_b$, we obtain the following observation equations (Option ①):

$$\begin{pmatrix} \delta y_{A_x} \\ \delta y_{A_y} \\ \delta y_{A_z} \end{pmatrix} = \begin{pmatrix} -2g\delta q_2 - \cancel{\delta a_{b_x}} \\ 2g\delta q_1 - \cancel{\delta a_{b_y}} \\ -\delta a_{b_z} \end{pmatrix}, \quad \begin{pmatrix} \delta y_{B_x} \\ \delta y_{B_y} \\ \delta y_{B_z} \end{pmatrix} = \begin{pmatrix} -2B_z\delta q_2 - \delta b_{b_x} \\ 2(B_z\delta q_1 - B_x\delta q_3) - \cancel{\delta b_{b_y}} \\ 2B_x\delta q_2 - \delta b_{b_z} \end{pmatrix}.$$

– Bias in δy_B: clearly, the bias δb_{b_y} in the measurement δy_{By} is unobservable regardless of our other choices. Indeed, we have that:

$$\delta_{bb_y} = -2\delta y_{B_y}(B_z\delta q_1 - B_x\delta q_3).$$

We cannot reconstruct both δq_3 and δb_{b_y} simultaneously from this one single relation, even in Case 2, when δa_{b_y} is not defined.

Option ①: However, we can apply the other two relations on δy_{B_x} in the observation model to find the values of δb_{b_x} and δb_{b_z}, provided that we do not consider any bias in δa_{b_x} in the first component of the acceleration. This leads to the following expressions:

$$\begin{cases} \delta b_{b_x} = -2B_z\delta q_2 - \delta y_{B_x} \\ \delta b_{b_z} = 2B_x\delta q_2 - \delta y_{B_z} \end{cases} \qquad [3.17]$$

where δq_2 is given by the formula:

$$\delta q_2 = -\frac{\delta y_{A_x}}{2g}.$$

– Bias in δy_A: continuing this reasoning, the bias δa_{b_y} cannot be observable when added to the measurement of the component δy_{A_y}, since this would require us to solve the following system:

$$\begin{cases} 2B_z\delta q_1 - 2B_x\delta q_3 = \delta y_{B_y} \\ 2g\delta q_1 - \delta a_{b_y} = \delta y_{A_y} \end{cases} \qquad [3.18]$$

We conclude that δa_{b_z} is the only estimable bias in the measurement δy_A.

CASE 2: maximize the number of imperfections modeled in the accelerometers

In this case, we can estimate two constant biases in y_A and one in y_B. If we repeat the analysis performed for Case 1, taking into account the natural unobservabilities of the system, we find (Option ②):

$$\begin{pmatrix} \delta y_{A_x} \\ \delta y_{A_y} \\ \delta y_{A_z} \end{pmatrix} = \begin{pmatrix} -2g\delta q_2 - \delta a_{b_x} \\ 2g\delta q_1 - \cancel{\delta a_{b_y}} \\ -\delta a_{b_z} \end{pmatrix}, \begin{pmatrix} \delta y_{B_x} \\ \delta y_{B_y} \\ \delta y_{B_z} \end{pmatrix} = \begin{pmatrix} -2B_z\delta q_2 - \cancel{\delta b_{b_x}} \\ 2(B_z\delta q_1 - B_x\delta q_3) - \cancel{\delta b_{b_y}} \\ 2B_x\delta q_2 - \delta b_{b_z} \end{pmatrix}$$

$$\text{or Option ③} \begin{pmatrix} -2B_z\delta q_2 - \delta b_{b_x} \\ 2(B_z\delta q_1 - B_x\delta q_3) - \cancel{\delta b_{b_y}} \\ 2B_x\delta q_2 - \cancel{\delta b_{b_z}} \end{pmatrix}.$$

– Bias in δy_A: we can observe δq_2, δq_2, δa_{b_x}, and δa_{b_z} from the equations:

$$\text{Option ②} \begin{cases} \delta q_1 &= -\frac{\delta y_{A_x}}{2g} \\ \delta q_2 &= -\frac{1}{2B_y}\delta y_{B_x} \\ \delta a_{b_z} &= -\delta y_{A_z} \\ \delta a_{b_x} &= -2g\delta q_2 - \delta y_{A_x} \end{cases} \quad \text{or Option ③} \begin{cases} \delta q_1 &= -\frac{\delta y_{A_x}}{2g} \\ \delta q_2 &= -\frac{\delta y_{B3}}{2B_y} \\ \delta a_{b_z} &= -\delta y_{A_z} \\ \delta a_{b_x} &= -2g\delta q_2 - \delta y_{A} \end{cases}$$

– Bias in δy_B: since the bias δb_{b_y} is unobservable, we can consider either an additive bias in δy_{B_x} or an additive bias in δy_{B_z}.

3.5.2. *Returning to our AHRS: integrating a description of measurement errors*

Extensive analysis is typically required when designing a high-performance AHRS to determine which measurement biases should be integrated into the estimation process. The imperfections must be chosen as carefully as possible. There are also various other choices to make, such as how to decouple the recalibrations of the prediction model. For example, it is often a good idea to only use the measurements from magnetometers to estimate the heading of the aircraft, since these readings are highly sensitive to environmental influences. Thus, whenever there are perturbations in the magnetic field measured by the sensor (which is very common in practice), only the heading angle will be affected. The roll and pitch components of the attitude will be unaffected by these frequent and often significant errors. Decoupling the estimation in this way can help to secure the flight of the mini-UAV. It has been shown that the same decoupling can instead be achieved by considering the vector product of the measurements from the accelerometers and the magnetometers rather than directly using the measurements from the magnetometers. This is done by using the relation $y_A \times y_B = q^{-1} * C * q := y_c$, where $C := A \times B$. Nevertheless, it is easy to see that $< A, C >= 0$, which implies that there is a mathematical dependency between these measurements and hence between any uncertainties modeled in them. This artificial operation therefore reduces the number of measured components from nine to eight, meaning that only five imperfections can then be evaluated by the estimation scheme to guarantee that the model remains fully observable. If we have already decided to measure three gyrometric biases (ω_m), then "only" two observable defect parameters remain. We can also draw another couple of interesting conclusions: (1) we can only estimate one single scalar imperfection in the accelerometer measurements if we wish to keep them independent of the magnetometer measurements (δq_2 gives a relation between them); (2) the estimator only depends on the longitudinal component of the Earth's magnetic field.

In light of this discussion, we shall choose to model the sensors as follows. The gyroscope measures $\omega_m = \omega + \omega_b$, where ω_b is a constant or slowly varying bias vector; the accelerometer measures $a_m = a_s \cdot a$, where $a_s > 0$ is a constant scale factor; finally, the magnetometer measures $y_B = b_s q^{-1} * B * q$, where $b_s > 0$ is also a constant or slowly varying scale factor. This gives the full model of the AHRS with an integrated description of the imperfections in the sensors:

$$\mathcal{M}_{ahrs}^{+}\begin{cases}\dot{q}=\frac{1}{2}q*(\omega_m-\omega_b) & \\ \dot{\omega}_b=0 & \text{(evolution)}\\ \dot{a}_s=0 & \\ \dot{b}_s=0 & \\ \begin{pmatrix}y_A\\ y_B\end{pmatrix}=\begin{pmatrix}a_s q^{-1}*A*q\\ b_s q^{-1}*B*q\end{pmatrix} & \text{(observation)}\end{cases}$$

where ω_m is the model input, assumed to be known. The value of a_m is compared against the predicted output y_A, and the magnetometer measurements are similarly compared against y_B. This model is fully observable, since every state variable can be reconstructed from the input and the measurements (ω_m, y_A, y_B) and their derivatives of all orders (and so the model is observable in the sense of definition 3.7). From this observation model, we can deduce the values of a_s and b_s from the formulas $a_s = \frac{1}{g}\|y_A\|$ and $b_s = \frac{1}{B_x g}\|y_B\|$. If a_s and b_s are known, the states associated with the quaternion q (i.e. q_1, q_2, q_3) may be deduced from the six rotational relations of the observation model. Finally, the evolution equations allow us to deduce the biases ω_b from the formula $\omega_b = \omega_m - 2q^{-1}\dot{q}$.

3.6. AHRS plus a GPS and a barometer: Inertial Navigation System

To avoid exceeding the budgetary constraints of our high-performance navigation system development project, we need to make do with MEMS-type inertial sensors. As we saw in section 1.3, in cases where inertial sensors are indispensable, we require access to additional instruments (such as a magnetometer) to improve our estimate of the heading of the aircraft. At the same time, the demands of mini-UAV navigation make it unavoidable for the state estimator to measure the velocity and the position of the vehicle by other exteroceptive sensors, so that the redundant measurements can be used to correct the navigation model presented in section 3.2.

One way to formulate this problem is to take advantage of a satellite positioning system (GPS) to generate another measurement of the velocity y_V and position y_X of the aircraft, both expressed in ground coordinates. When considering the "true" inertial navigation problem in section 3.3, we showed that the error in the vertical velocity diverges linearly by performing a detectability study on the first-order linearized model. By introducing another measurement system such as GPS to obtain additional vertical information, among other things, we can prevent this divergence. The GPS measurement is itself far from perfect or self-sufficient, due to its variable accuracy. Furthermore, loss of GPS signal is not uncommon in some flight conditions. Accordingly, it is sensible and perhaps even necessary to perform wide-scope information-merging whenever some kind of redundancy can be implemented within these types of aircraft cell. For the vertical position (or altitude), baroaltimeters offer an alternative source of measurements, giving extra information about the altitude $y_h =< X, e_3 >$. Combining this with the position measurement y_X delivered by the GPS improves the accuracy of our estimate of the aircraft's vertical position. Our nonlinear state estimator shall therefore use a GPS and a pressure sensor, as well as the three triaxial sensors of the AHRS. In summary, the Inertial Navigation System (INS) is based on the following 16 scalar measurements:

– six datapoints generated by the GPS, namely the measurements of the velocity and position vectors of the aircraft, respectively, denoted by $y_X = X \in \mathbb{R}^3$ and $y_V = V \in \mathbb{R}^3$;

– the altitude measurement $y_h =< X, e_3 >$ generated by a pressure sensor (or barometer), used in combination with the GPS measurement to improve the estimate of the vertical position.

Starting from the very general model $\mathcal{M}_s$ formulated in section 3.2, we can now derive the equations of the complete navigation system. The new model $\mathcal{M}_{ins}$ may be stated as follows:

$$\mathcal{M}_{ins} \begin{cases} \dot{q} = \frac{1}{2} q * \omega_m \\ \dot{V} = A + q * a_m * q^{-1} \qquad \text{(evolution)} \\ \dot{X} = V \\ \begin{pmatrix} y_V \\ y_X \\ y_h \\ y_B \end{pmatrix} = \begin{pmatrix} V \\ X \\ < X, e_3 > \\ q^{-1} * B * q \end{pmatrix} \qquad \text{(observation)} \end{cases}$$

INS is used to describe any system that estimates the orientation, velocity and position from low-accuracy sensors and a GPS receiver.

3.6.1. *Observability of the imperfections in the measurements used by the INS*

As we saw earlier in section 3.5.1, we have some freedom in the choice of how many and which imperfections should be modeled in order to be re-estimated. The only constraint is that the system of equations $\mathcal{M}_{ins}$ must remain fully observable. Linearizing the model $\mathcal{M}_{ins}$ about the equilibrium point $(\bar{q} = (1\ 0\ 0\ 0)^T \quad \omega = \vec{0} \quad \dot{V} = \vec{0} \quad \bar{X})^T$ gives the following relations:

$$\begin{cases} \delta\dot{q} &= 0.5\delta\omega \\ \delta\dot{V} &= 2A \wedge \delta q + \delta a \\ \delta\dot{X} &= \delta V \end{cases}$$

$$\begin{cases} \delta X_x &= \delta y_{Xx} \\ \delta X_y &= \delta y_{Xy} \\ \delta X_z &= \delta y_{Xz} \end{cases}, \begin{cases} \delta V_x &= \delta y_{Vx} \\ \delta V_y &= \delta y_{Vy} \\ \delta V_z &= \delta y_{Vz} \end{cases}, \delta X_3 = \delta y_h,$$

$$\begin{cases} -2B_z\delta q_2 &= \delta y_{Bx} \\ 2B_z\delta q_1 - 2B_x\delta q_3 &= \delta y_{By} \\ 2B_x\delta q_2 &= \delta y_{Bz} \end{cases}$$

We therefore have a set of six measured inputs, assumed to be known, and 10 measurements, from which we can deduce a set of model outputs. We need to estimate a total of nine state variables in order to fully determine the orientation, velocity and position of the aircraft. We can therefore re-estimate up to $16 - 9 = 7$ of the imperfections in the various sensors. This number can even be increased to 10 if we observe that the position may be deduced by integrating the velocity directly over time (this is equivalent to adding three additive bias terms to the position measurement generated by the GPS). It is useful to note that the state δX_z can be determined either from the GPS measurement δy_{Xz} or the barometer measurement δy_h, since the instruments are redundant. Therefore, we can recalibrate any bias in the barometric measurement by using the GPS measurement. This hybrid approach allows us to achieve better accuracy in the estimate of the vertical position of the mini-UAV and correct any altitude errors that may for example have arisen by a sudden change in the atmospheric pressure or the wind.

The degrees of freedom are fixed as follows: based on the four triaxial sensors and a pressure sensor, which give a total of 16 scalar measurements, the gyroscope measures $\omega_m = \omega + \omega_b$, where ω_b is a constant or slowly changing vector bias term; the accelerometer measures $a_m = a_s \cdot a$, where $a_s > 0$ is a nearly constant scale factor; the barometer is assigned a scalar measurement bias h_b. This gives the full model of the INS with an integrated description of the imperfections in the on-board sensors:

$$
\mathcal{M}^+_{ins} \begin{cases} \dot{q} = \frac{1}{2} q * (\omega_m - \omega_b) \\ \dot{V} = A + \frac{1}{a_s} q * a_m * q^{-1} & \text{(evolution)} \\ \dot{X} = V \\ \dot{\omega}_b = 0 \\ \dot{a}_s = 0 \\ \dot{h}_b = 0 \\ \\ \begin{pmatrix} y_V \\ y_X \\ y_h \\ y_B \end{pmatrix} = \begin{pmatrix} V \\ X \\ < X, e_3 > -h_b \\ q^{-1} * B * q \end{pmatrix} & \text{(observation),} \end{cases}
$$

where ω_m and a_m are imperfect and noisy inputs, assumed to be known, and $[V^T \; X^T \; B^T]^T$ are the measured outputs, also imperfect and noisy. Just like the model $\mathcal{M}^+_{ahrs}$, this system is observable, since every state variable can be reconstructed from the inputs and outputs $\omega_m, a_m, y_V, y_X, y_h, y_B$ and their derivatives. For instance, from these equations, we can deduce that $a_s = \frac{\|a_m\|}{\|\dot{y}_V - A\|}$ and $\frac{a_m}{\|a_m\|} = q^{-1} * \frac{\dot{y}_V - A}{\|\dot{y}_V - A\|} * q$. If we know the rotation formulas that relate q to ($\dot{y}_V$-A) and B, we can find an expression for q in the sense of definition 3.7. The evolution equations of the system then allow us to deduce the biases in the gyrometers, which satisfy $\omega_b = \omega_m - 2q^{-1}\dot{q}$, and the bias in the barometer, which satisfies $h_b =< y_X, e_3 > -y_h$.

4

The IUKF and π-IUKF Algorithms

4.1. Preliminary remarks

In Chapter 3, we saw that there can be multiple degrees of freedom when establishing an accurate dynamic model for inertial navigation, allowing us to fix a certain number of modeling assumptions (for example describing some of the imperfections in the measurement process) without losing the observability of the system state estimation problem. This chapter develops a set of original methodological principles that are used to establish two nonlinear estimation algorithms for the models $\mathcal{M}_{ahrs}$ and $\mathcal{M}_{ins}$. By combining invariant observer theory and the principles of so-called unscented Kalman filtering (see Chapter 2), these algorithms can compute a specific correction term for any prediction derived from a nonlinear state representation in such a way that the dynamics of the observer thus constructed satisfy the same symmetry properties as the system itself. In a similar spirit to research from a few years ago on the Invariant Extended Kalman Filter (IEKF) algorithm, the correction gains of this estimator, which are specifically designed to be invariant, may be deduced by performing the same computational steps as UKF-type filtering (either in factorized or non-factorized form). However, before we can integrate the procedure for computing the correction gains (an algorithm borrowed from unscented Kalman filtering) with invariant observer theory, a series of methodological developments are required, as described in this chapter. Our research ultimately led us to formulate two variants of the UKF algorithm, which we named Invariant Unscented Kalman Filter (IUKF) and π-IUKF [CON 13]

[CON 14]. Factorized Square Root (SR) forms of both methods can easily be derived from the original formulations.

In particular, our study of the (SR)-IUKF and π-(SR)-IUKF filters shows that the principles of Sigma-Point Kalman Filtering (SPKF) may be used to construct invariant filters characterized by a large domain of convergence that are capable of estimating the state of dynamic nonlinear systems. At the same time, using a UKF technique to compute the gain terms completely avoids one of the obstacles associated with the IEKF approach, since we are not forced to explicitly linearize the dynamics of the invariant state error in order to identify the matrices $\mathbf{A}(I)$ and $\mathbf{C}(I)$, which depend on the invariants of the problem and are needed if we wish to apply the principles of extended Kalman filtering. Although the dynamics can be differentiated analytically in some special cases (typically whenever the motion of the system satisfies a system of kinetic formulas), which effectively eliminates this obstacle in these cases, the theoretical approach of this book is motivated by the ambition of being generally applicable to any type of dynamic system that might be encountered in future. In particular, this includes cases where the inherent complexity of the model of the system dynamics prevents the linearization from being performed analytically. For example, numerical representations presented as data tables describing the aerodynamics and propulsive behavior of fixed-wing mini-UAVs, can help us to characterize the flight dynamics of such aircraft more comprehensively (by providing a rich *a priori* knowledge base for the system). However, the non-differentiable and often extremely complex nature of these fundamental "building blocks" of the model undermines any purely constructive approach to establishing the linear differentiation equations satisfied by the invariant state error of the estimation problem. Compared with IEKF, the approaches and techniques proposed here therefore seem more applicable to arbitrary dynamic models, since they only rely on knowledge of the composite group transformation $\phi_g = \{\varphi_g, \psi_g, \rho_g\}$.

4.2. Organization of this chapter

This chapter is divided into six parts:

– Section 4.3 reviews a series of mathematical results from differential geometry. These results offer some insight into the concepts of symmetry and invariance in the context of dynamics systems, which are heavily exploited throughout the rest of the chapter.

– Sections 4.4 and 4.5 construct two invariant estimators that provide a promising algorithmic solution to the AHRS and INS inertial navigation problems (see Chapter 3). The filter equations thus obtained yield an explicit description of the group transformations defined for the evolution and observation models of $\mathcal{M}_{ahrs}$ and $\mathcal{M}_{ins}$. This allows us to establish a suitable mathematical formulation of the corresponding estimation problems. The state, control, and output transformations formalize the symmetries and invariance properties of the dynamics and the observation process of the system, which enables us to define: (1) an invariant vector field that fixes the geometric setting of the estimation process (by applying E. Cartan's method of moving frames); (2) a suitable (invariant) output error that we may use to directly estimate the state. As we shall see, this allows us to fix the dynamics of the estimator and the computed gain terms (either analytically or with an EKF) by linearizing, or in some cases approximating up to the second order, the set of differential equations governing the evolution of the invariant state estimation error over time. The decision to use an invariant state estimation error greatly simplifies our analysis of the convergence of the estimating filter. Given a certain set of target dynamics, the linearization operation enables us to compute numerical values for the correction terms that endow the filter with these dynamics and simultaneously guarantees that the domain of convergence of the invariant state estimation error is sufficiently large. In passing, note that the differential equation satisfied by the dynamics of the invariant state estimation error is of a certain form; in fact, it is an autonomous equation and therefore only depends on the error itself, as well as the fundamental invariants of the system. This guarantees that the domain of convergence of the error will be sufficiently large. Nevertheless, the linearization procedure mentioned above can be difficult to execute in practice, especially for dynamic systems with complex models. Ultimately, this means that it may be preferable to choose a less constructive but more systematic approach to calculating the correction gain terms.

– Before presenting the two algorithms that avoid this linearization step, allowing us to construct invariant estimating filters to estimate the state of dynamic systems, section 4.6 briefly presents each of the calculation steps of the (SR)-UKF algorithm. This algorithm provides a basis for the discussion throughout the rest of the chapter. This section also lists and discusses the known advantages and disadvantages of this estimation technique.

– Accordingly, section 4.7 is dedicated to our first reformulation of the so-called unscented Kalman filter for an invariant setting. As we shall see, the notions of state errors and invariant outputs play a predominant role in redefining the covariance matrices used by sigma-points Kalman filtering. Similarly, the moving frame, defined and calculated about the predicted state, appears directly in the equations of the estimating filter, in a manner analogous to IEKF. We chose to call this algorithm IUKF. A factorized square-root formulation can easily be derived (in the same way as classical UKF). A comprehensive set of results for the AHRS estimation problem generated from simulated noisy data are then presented to demonstrate the well-foundedness of this algorithm, as well as its potential benefits in both theoretical and practical contexts.

– Finally, after this first attempt to unite the principles of sigma-points Kalman filtering and the theory of invariant observers, the sixth part of this chapter presents a factorized version of the π-IUKF algorithm. In particular, we shall see that there is another way to recover the invariance properties of the algorithm presented in section 4.7 by applying different modifications to the standard Square-Root Unscented Kalman Filter (SR-UKF) algorithm from section 4.6. By defining a fourth nonlinear transformation π, which is applied to the observation equations of the model during the output prediction step, we can deduce an invariant estimator more easily without needing to redefine the set of covariance matrices, as would be required by standard IUKF (see section 4.7). This additional transformation can be interpreted as a *compatibility condition* between the dynamics of the system (which can be left or right-invariant) and the specific dynamics of the estimator (which can also satisfy the same symmetry properties as the system). Recall that observers are always based on a copy of the dynamics of the system. The π-IUKF algorithm is therefore itself closer to the original version of SR-UKF, since it requires very few modifications. The analytic expressions that define the mapping π are given in this section for both the AHRS and the INS. Chapter 5 will focus on performing an in-depth evaluation of this second algorithm.

4.3. Results from differential geometry: symmetries and invariant/equivariant systems

The word *symmetry* (first used in 1529), derived from the Greek word *symmetros* meaning "of same measure," was originally an architectural term which referred to the parts of a building that contribute to its beauty as a whole. In the 17th century, the meaning of the word was extended to incorporate the idea of regularity among parts of a whole and used to describe certain pieces of art and music. In the 18th century, Diderot proposed a specialized definition of the word to refer to identical elements that are regularly distributed either side of an axis. Symmetry would later become a specific scientific and geometric term describing the similarity of two figures with respect to a plane around the year 1870; thus, the concepts of an axis or plane of symmetry were born. Today, the notion of *symmetry* can be understood as a generalization of two concepts:

– the *redundancy* of irreducible parts of a whole;

– and the *invariance properties* of an object that cause it to be left unchanged by a set of transformations.

Historically, physicists and mathematicians from a variety of fields have devoted extensive time and energy to developing algebraic and numerical methods that would allow symmetry groups to be systematically identified. These methods are based on the concepts of invariance, equivalence and indiscernability (see [OLV 03] and [OLV 08]), which are characteristic properties of the vast majority of the laws of physics. More precisely, these research techniques work by identifying group transformations that leave the (differential) equations describing the (dynamics of) the system unchanged. Here, we briefly review a collection of results from recent research that can be applied to identify and exploit the symmetry groups of dynamic systems. First, consider the nonlinear state representation defined as follows:

$$\begin{cases} \dot{\mathbf{x}} = f(\mathbf{x}, \mathbf{u}) \\ \mathbf{y} = h(\mathbf{x}, \mathbf{u}) \end{cases} \qquad [4.1]$$

In equation [4.1], the state vector (respectively, vector of inputs or vector of outputs) belongs to the open set $\mathcal{X} \subset \mathbb{R}^n$ (respectively, $\mathcal{U} \subset \mathbb{R}^m$, $\mathcal{Y} \subset \mathbb{R}^p$, $p \leq n$). Consider also a composite group transformation $\phi_{g \in G}$ acting on the Cartesian product of the sets $\mathcal{X} \times \mathcal{U} \times \mathcal{Y}$ defined by:

$$\begin{aligned} &\phi_g : G \times (\mathcal{X} \times \mathcal{U} \times \mathcal{Y}) \longrightarrow (\mathcal{X} \times \mathcal{U} \times \mathcal{Y}) \\ &(g, \mathbf{x}, \mathbf{u}, \mathbf{y}) \longmapsto \phi_g(\mathbf{x}, \mathbf{u}, \mathbf{y}) = (\varphi_g(\mathbf{x}), \psi_g(\mathbf{u}), \rho_g(\mathbf{y})) = (\mathbf{X}, \mathbf{U}, \mathbf{Y}), \end{aligned}$$

where G is a group equipped with a Lie algebra; in other words, G may be viewed as a finite-dimensional differential manifold on which the group operations of inversion and multiplication are differentiable mappings. Additionally, let $(\varphi_g, \psi_g, \rho_g)$ be three diffeomorphisms parametrized by $g \in G$. In practice, finding the symmetries of a dynamic system involves finding three local transformations defined on three open sets that act separately on the variables of the system model (i.e. distinct actions on the state, the input and the output) such that the evolution equations $d\mathbf{x}/dt = f(\mathbf{x}, \mathbf{u})$ and observation equations $\mathbf{y} = h(\mathbf{x}, \mathbf{u})$ are fixed by each transformation (i.e. $\dot{\mathbf{X}} = f(\mathbf{X}, \mathbf{U})$ and $\mathbf{Y} = h(\mathbf{X}, \mathbf{U})$). The dynamics of the system are said to be G-invariant if $\exists(\varphi_g, \psi_g)_{g \in G}$, $\forall(g, \mathbf{x}, \mathbf{u}) \in G \times \mathcal{X} \times \mathcal{U}$, $f(\varphi_g(\mathbf{x}), \psi_g(\mathbf{u})) = D\varphi_g(\mathbf{x}) \cdot f(\mathbf{x}, \mathbf{u})$, and are said to be G-equivariant if, in addition to the previous condition, $\exists(\rho_g)_{g \in G}$ acting on $\mathcal{Y}$, $\forall(g, \mathbf{x}, \mathbf{u}) \in G \times \mathcal{X} \times \mathcal{U}$, $h(\varphi_g(\mathbf{x}), \psi_g(\mathbf{u})) = \rho_g(h(\mathbf{x}, \mathbf{u}))$. In other words, every differential relation (f) and static relation (h) remains explicitly identical. In the context of mini-UAV applications, we can give physical interpretations of the invariance properties of the kinematic relations. Each invariance corresponds to a symmetry of the dynamics of the motion of the aircraft that is independent of the reference system (aircraft or ground coordinates) in which it is expressed (symmetries and invariance under the action of the group of translations/rotations).

4.4. Invariant observers – AHRS/INS

4.4.1. *AHRS: model summary*

By following the construction method presented in Chapter 2, we will define a very general form of invariant observer theory for the AHRS estimation problem (see Figure 4.1). The model of the AHRS problem introduced in the previous chapter can be briefly restated as follows. The nonlinear state representation of the model $\mathcal{M}^{+}_{ahrs}$ may be compactly written as $\dot{\mathbf{x}} = f(\mathbf{x}, \mathbf{u})$ and $\mathbf{y} = h(\mathbf{x}, \mathbf{u})$, where $\mathbf{x} = (q^T \; \omega_b^T \; a_s \; b_s)^T$, $\mathbf{u} = (\omega_m)$ and

$\mathbf{y} = (y_A^T \, y_B^T)^T$ are defined by:

$$\mathcal{M}_{ahrs}^{+} \begin{cases} \dot{q} = \frac{1}{2} q * (\omega_m - \omega_b) \\ \dot{\omega}_b = 0 & \text{(evolution)} \\ \dot{a}_s = 0 \\ \dot{b}_s = 0 \\ \begin{pmatrix} y_A \\ y_B \end{pmatrix} = \begin{pmatrix} a_s q^{-1} * A * q \\ b_s q^{-1} * B * q \end{pmatrix} & \text{(observation)} \end{cases} \qquad [4.2]$$

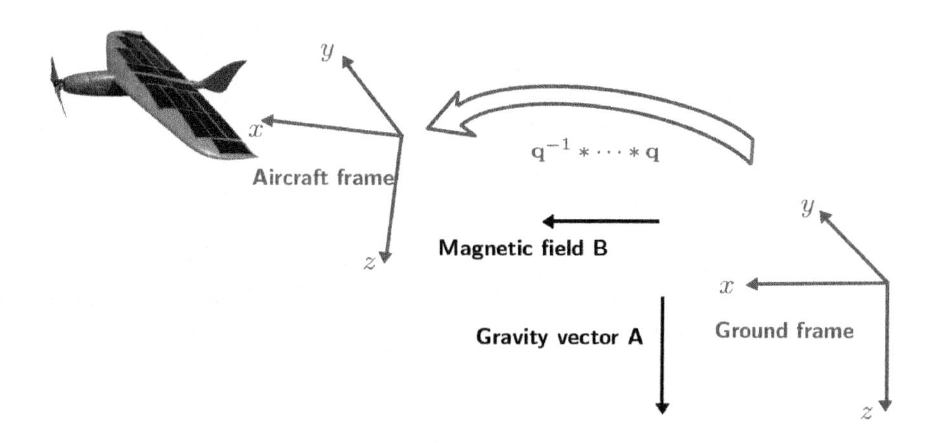

Figure 4.1. *AHRS – Illustration of the transition between ground coordinates and aircraft coordinates. For a color version of this figure, see www.iste.co.uk/condomines/kalman.zip*

4.4.2. *Invariance of the AHRS model equations*

Given the equations of the model $\mathcal{M}_{ahrs}^{+}$ defined by equation [4.2] and the Lie group $G = \mathbb{H}_1 \times \mathbb{R}^5$ (where $\mathbb{H}_1$ is the Lie algebra of quaternions of norm one) acting on the entire state space on which $\mathcal{M}_{ahrs}^{+}$ is defined (we say that the group is a free group, see Appendix), the local transformations stated below allow us to verify that the dynamics of the system are indeed G-invariant and G-equivariant. For all $g = (q_0^T \, \omega_0^T \, a_0 \, c_0)^T \in G$, we have that:

$$1)\ \varphi_g\left(\mathbf{x} = \left(q^T\ \omega_b^T\ a_s\ b_s\right)^T\right) = \begin{pmatrix} q * q_0 \\ q_0^{-1} * \omega_b * q_0 + \omega_0 \\ a_s \cdot a_0 \\ b_s \cdot b_0 \end{pmatrix};$$

$$2)\ \psi_g(\mathbf{u} = (\omega_m)) = q_0^{-1} * \omega_m * q_0 + \omega_0;$$

$$3)\ \rho_g\left(\mathbf{y} = \left(y_A^T\ y_B^T\right)^T\right) = \begin{pmatrix} a_0 \cdot q_0^{-1} * y_A * q_0 \\ b_0 \cdot q_0^{-1} * y_B * q_0 \end{pmatrix}.$$

Geometrically, each transformation is equivalent to either a rotation, a translation or the composition of a rotation and a translation. These transformations act as the links between the parameters $\mathbf{x}$, $\mathbf{u}$ and $\mathbf{y}$ expressed in ground coordinates and the same parameters in aircraft coordinates, illustrating the Galilean invariance structure of the estimation problem. For a computational proof that the dynamics of the system are invariant, consider the new parameters $Q = q * q_0$, $\Omega_b = q_0^{-1} * \omega_b * q_0 + \omega_0$, $\Omega_m = q_0^{-1} * \omega_m * q_0 + \omega_0$, $\mathcal{A}_s = a_s \cdot a_0$ and $\mathcal{B}_s = b_s \cdot b_0$. With these parameters, it is straightforward to show that the system $\mathcal{M}_{ahrs}^{+}$ is indeed invariant under the group transformations defined above as follows:

$$\begin{aligned} \dot{Q} &= \overbrace{(q * q_0)}^{\cdot} = \left(\frac{1}{2} q * (\omega_m - \omega_b)\right) * q_0 = \frac{1}{2}(q * (\omega_m - \omega_b)) * q_0 \\ &= \frac{1}{2}(q * \omega_m - q * \omega_b) * q_0 = \frac{1}{2}(q * \omega_m * q_0 - q * \omega_b * q_0) \\ &= \frac{1}{2}(q * q_0 * q_0^{-1} * \omega_m * q_0 - q * q_0 * q_0^{-1} * \omega_b * q_0) \\ &= \frac{1}{2} Q * ((q_0^{-1} * \omega_m * q_0 + \omega_0) - (q_0^{-1} * \omega_b * q_0 + \omega_0)) \\ &= \frac{1}{2} Q * (\Omega_m - \Omega_b), \end{aligned}$$

$$\dot{\Omega}_b = \overbrace{(q_0^{-1} * \omega_b * q_0 + \omega_0)}^{\cdot} = q_0^{-1} * \dot{\omega}_b * q_0 = 0,$$

$$\dot{\mathcal{A}}_s = \overbrace{(a_s \cdot a_0)}^{\cdot} = \dot{a}_s \cdot a_0 = 0,$$

$$\dot{\mathcal{B}}_s = \overbrace{(b_s \cdot b_0)}^{\cdot} = \dot{b}_s \cdot b_0 = 0.$$

After computing the derivative of the image of each component of the state vector under the mapping φ_g with respect to time, we recover the same differential equations as the original system. The dynamics of the AHRS model are therefore G-invariant. Similarly, we can establish that the static equations of the observation model are G-equivariant computationally from the fact that they remain explicitly identical after applying the local transformation ρ_g defined earlier. The term $\rho_g(\mathbf{y})$ can be expressed as the image under h of the various new state and control variables introduced above. Thus:

$$\begin{aligned}\rho_g(\mathbf{y}) = \begin{pmatrix} a_0 \cdot q_0^{-1} * y_A * q_0 \\ b_0 \cdot q_0^{-1} * y_B * q_0 \end{pmatrix} &= \begin{pmatrix} a_0 \cdot q_0^{-1} * (a_s \cdot q^{-1} * A * q) * q_0 \\ b_0 \cdot q_0^{-1} * (b_s \cdot q^{-1} * B * q) * q_0 \end{pmatrix} \\ &= \begin{pmatrix} a_s \cdot a_0 \cdot \overbrace{(q * q_0)^{-1}}^{Q^{-1}} * A * \overbrace{(q * q_0)}^{Q} \\ b_s \cdot b_0 \cdot Q^{-1} * B * Q \end{pmatrix} \\ &= h\left(\varphi_g(\mathbf{x}), \psi_g(\mathbf{u})\right).\end{aligned}$$

4.4.3. *Constructing an output error and an invariant frame*

If we consider the special case $g = \mathbf{x}^{-1} = ((q^{-1})^T \ (-q * \omega_b * q^{-1})^T \ a_s^{-1} \ b_s^{-1})^T$ in the above calculations, we find:

$$\varphi_{\mathbf{x}^{-1}}(\mathbf{x}) = \begin{pmatrix} q * q^{-1} \\ q * \omega_b * q^{-1} - q * \omega_b * q^{-1} \\ a_s \cdot a_s^{-1} \\ b_s \cdot b_s^{-1} \end{pmatrix} = \begin{pmatrix} \mathbf{1} \\ \vec{0} \\ 1 \\ 1 \end{pmatrix} = e. \qquad [4.3]$$

In equation [4.3], e denotes the identity element of the local transformation φ_g. Each component of the vector $(\mathbf{1} \ \vec{0} \ 1 \ 1)^T$ is the identity element of a subspace equipped with a specific operation:

– $\mathbf{1} = (1 \ 0 \ 0 \ 0)^T$ is the identity element of the quaternions under the Hamiltonian product ($*$);

– $\vec{0} = \vec{0}_{\mathbb{R}^3}$ is the identity element of spatial translations, i.e. the identity element of $\mathbb{R}^3$ under vector addition ($+$);

– the constant 1 is the identity element of the scalar real numbers under multiplication ($\cdot$).

Equation [4.3] defines a vector of invariants, where $\mathbf{x}^{-1} \in G$ parametrizes the Lie group that yields these invariants. These equations are known as the normalization equations (see Appendix). They allow us to construct an invariant output error E as well as an invariant frame. We shall use the following notation for convenience:

$$\gamma(\mathbf{x}) = \begin{pmatrix} q^{-1} \\ -q * \omega_b * q^{-1} \\ a_s^{-1} \\ b_s^{-1} \end{pmatrix}. \quad [4.4]$$

With this notation, the general form of an invariant observer can be written as follows:

$$\dot{\hat{\mathbf{x}}} = \overbrace{f(\hat{\mathbf{x}}, \mathbf{u})}^{\text{G-invariant prediction}} + \overbrace{\sum_{i=1}^{n} \mathbf{K}^{(i)} \left(\psi_{\hat{\mathbf{x}}^{-1}}, \mathsf{E} := \left[h\left(e, \psi_{\hat{\mathbf{x}}^{-1}}(\mathbf{u})\right) - \rho_{\hat{\mathbf{x}}^{-1}}(\mathbf{y})\right]\right) w_i(\hat{\mathbf{x}})}^{\text{invariant correction terms}}. \quad [4.5]$$

In equation [4.5], $\hat{\mathbf{x}} \in \mathbb{R}^n$ is the estimated state vector. It is easy to recognize that this gives a standard expression for a nonlinear state estimator in which the prediction model (with G-invariant dynamics and G-equivariant outputs) is summed with an invariant correction term. The key idea when constructing an invariant estimator is to define an additive correction term based on nonlinear gain terms $\mathbf{K}^{(i)}$ that depend nonlinearly on the system invariants $(e, \psi_{\hat{\mathbf{x}}^{-1}}, \rho_{\hat{\mathbf{x}}^{-1}})$. This equips the dynamics of the estimator with the same G-invariance properties as the observed system. Equation [4.5] gives a very general expression for the invariant output error E. In the case of the AHRS, it is relatively straightforward to fill in the details from the various model equations. Thus:

$$\begin{aligned} \mathsf{E} &= h\left(e = \varphi_{\hat{\mathbf{x}}^{-1}}(\hat{\mathbf{x}}), \psi_{\hat{\mathbf{x}}^{-1}}(\mathbf{u})\right) - \rho_{\hat{\mathbf{x}}^{-1}}(\mathbf{y}) \\ &= \begin{pmatrix} A - \hat{a}_s^{-1}\hat{q} * y_A * \hat{q}^{-1} \\ B - \hat{b}_s^{-1}\hat{q} * y_B * \hat{q}^{-1} \end{pmatrix} = \begin{pmatrix} \mathsf{E}_A \\ \mathsf{E}_B \end{pmatrix}. \end{aligned} \quad [4.6]$$

The nonlinear output error in equation [4.6] can easily be shown to be invariant. As an example, consider the output error E_A associated with the accelerometric outputs ($\forall g = (q_0^T \ \omega_0^T \ a_0 \ c_0)^T \in G$):

$$\begin{aligned}
\mathsf{E}_A(\varphi_g(\hat{\mathbf{x}}), \rho_g(y_A)) &= \mathsf{E}_A(\overbrace{\hat{q} * q_0, q_0^{-1} * \hat{\omega}_b * q_0 + \omega_0, a_0\hat{a}_s, c_0\hat{c}_s}^{\varphi_g(\hat{\mathbf{x}})}, \overbrace{a_0 q_0^{-1} * y_A * q_0}^{\rho_g(y_A)}) \\
&= A - (a_0 \cdot \hat{a}_s)^{-1}(\hat{q} * q_0) * (a_0 q_0^{-1} * y_A * q_0) * (\hat{q} * q_0)^{-1} \\
&= A - \hat{a}_s^{-1}\hat{q} * y_A * \hat{q}^{-1} \\
&= \mathsf{E}_A(\hat{q}, \hat{\omega}_b, \hat{a}_s, \hat{c}_s, y_A) \\
&= \mathsf{E}_A(\hat{\mathbf{x}}, y_A).
\end{aligned}$$

The vectors $w_i(\hat{\mathbf{x}})$ define an invariant frame. In other words, for each $\hat{\mathbf{x}} \in \mathcal{X}$, the set $\{w_1(\hat{\mathbf{x}}) \ \ldots \ w_n(\hat{\mathbf{x}})\}$ consists of the invariant correction terms projected onto each component of $f(\hat{\mathbf{x}}, \mathbf{u})$, i.e. onto the tangent space generated by the functional f at $\hat{\mathbf{x}}$ (see Appendix). In our case, if we consider the standard basis of $\mathbb{R}^3$ with the vectors $(e_i)_{i \in [1;3]}$, then we can obtain expressions for the vectors of the invariant frame (also called the natural basis of the tangent space) by evaluating each of the eight vector equations at the point $\hat{\mathbf{x}} = (\hat{q}^T \ \hat{\omega}_b^T \ \hat{a}_s \ \hat{b}_s)$:

$$\overbrace{\left[D\varphi_{\gamma(\mathbf{x})} \begin{pmatrix} q \\ \omega_b \\ a_s \\ b_s \end{pmatrix} \right]}^{\text{Jacobian matrix of the geometric transformation } \varphi_g(\mathbf{x})} \cdot w(\mathbf{x}) = \begin{pmatrix} 0 \\ e_i \\ \hline 0 \\ 0 \\ 0 \\ \hline 0 \\ \hline 0 \end{pmatrix}_{i \in [1;3]}, \begin{pmatrix} 0 \\ 0 \\ 0 \\ 0 \\ \hline e_i \\ \hline 0 \\ \hline 0 \end{pmatrix}_{i \in [1;3]}, \begin{pmatrix} 0 \\ 0 \\ 0 \\ 0 \\ \hline 0 \\ 0 \\ 0 \\ \hline 1 \\ \hline 0 \end{pmatrix} \text{ and } \begin{pmatrix} 0 \\ 0 \\ 0 \\ 0 \\ \hline 0 \\ 0 \\ 0 \\ \hline 0 \\ \hline 1 \end{pmatrix}. \qquad [4.7]$$

After inverting the Jacobian matrix $D\varphi_{\gamma(\mathbf{x})}(\mathbf{x})$ in equation [4.7] by applying the local inversion theorem to deduce that $[D\varphi_{\gamma(\mathbf{x})}(\mathbf{x})]^{-1} = D\varphi_{\gamma^{-1}(\mathbf{x})}(\mathbf{x})$, we obtain the following invariant vector field:

$$\begin{pmatrix} w_i^q(\mathbf{x}) \\ w_i^{\omega_b}(\mathbf{x}) \\ w^{a_s}(\mathbf{x}) \\ w^{b_s}(\mathbf{x}) \end{pmatrix} = \left\{ \begin{pmatrix} 0 \\ e_i * q \\ \hline 0 \\ 0 \\ 0 \\ \hline 0 \\ \hline \\ 0 \end{pmatrix} \right\}_{i\in[1;3]}, \left\{ \begin{pmatrix} 0 \\ 0 \\ 0 \\ 0 \\ \hline q^{-1} * e_i * q \\ \hline 0 \\ \hline \\ 0 \end{pmatrix} \right\}_{i\in[1;3]}, \begin{pmatrix} 0 \\ 0 \\ 0 \\ 0 \\ \hline 0 \\ 0 \\ 0 \\ \hline a_s \\ \hline 0 \end{pmatrix} \text{ and } \begin{pmatrix} 0 \\ 0 \\ 0 \\ 0 \\ \hline 0 \\ 0 \\ 0 \\ \hline 0 \\ \hline b_s \end{pmatrix}. \qquad [4.8]$$

4.4.4. *Formulating the invariant observer for a low-budget AHRS*

To formulate the invariant observer, we need a few hypotheses in order to establish analytic expressions and values for the gain terms. If no error is observed in the estimated system outputs, then the prediction equations of the model do not require any corrections. Therefore, the constant term of the first-order Taylor expansion of the gain terms in equation [4.5] about an invariant output error of zero is zero. Thus, after linearizing the gain terms, which have an *a priori* nonlinear dependency on the invariants of the problem, we can write:

$$\forall i \in [\![1\,;n]\!], \begin{cases} \mathbf{K}^{(i)}(\psi_{\hat{\mathbf{x}}^{-1}}(\mathbf{u}), \mathsf{E}=0) = 0 \\ \forall \mathsf{E} \neq 0,\ \mathbf{K}^{(i)}(\psi_{\hat{\mathbf{x}}^{-1}}(\mathbf{u}), \mathsf{E}) \simeq \bar{\mathbf{K}}^{(i)}(\psi_{\hat{\mathbf{x}}^{-1}}(\mathbf{u}), \mathsf{E}) \cdot \mathsf{E} \end{cases} \qquad [4.9]$$

From this, we can immediately deduce the following linear approximation of equation [4.5]:

$$\dot{\hat{\mathbf{x}}} = f(\hat{\mathbf{x}}, \mathbf{u}) + \sum_{i=1}^{n} \bar{\mathbf{K}}^{(i)}(\psi_{\hat{\mathbf{x}}^{-1}}(\mathbf{u}), \mathsf{E}) \cdot \mathsf{E} \cdot w_i(\hat{\mathbf{x}}). \qquad [4.10]$$

After combining all of these results, equation [4.10] can be rewritten as follows in the special case of an AHRS[1]:

$$\mathcal{O}_{\mathcal{M}^+_{ahrs}} \begin{cases} \dot{\hat{q}} = \frac{1}{2}\hat{q} * (\omega_m - \hat{\omega}_b) + \sum_{i=1}^{3}(\bar{\mathbf{K}}^{(i)}_{A\to q} \cdot \mathrm{E}_A + \bar{\mathbf{K}}^{(i)}_{B\to q} \cdot \mathrm{E}_B) \cdot e_i * \hat{q} \\ \dot{\hat{\omega}}_b = \hat{q}^{-1} * \left(\sum_{i=1}^{3} \bar{\mathbf{K}}^{(i)}_{A\to\omega_b} \cdot \mathrm{E}_A + \bar{\mathbf{K}}^{(i)}_{B\to\omega_b} \cdot \mathrm{E}_B \right) * \hat{q} \\ \dot{\hat{a}}_s = \hat{a}_s \cdot (\bar{\mathbf{K}}^{(1)}_{A\to a_s} \cdot \mathrm{E}_A + \bar{\mathbf{K}}^{(1)}_{B\to a_s} \cdot \mathrm{E}_B) \\ \dot{\hat{b}}_s = \hat{b}_s \cdot (\bar{\mathbf{K}}^{(1)}_{A\to b_s} \cdot \mathrm{E}_A + \bar{\mathbf{K}}^{(1)}_{B\to b_s} \cdot \mathrm{E}_B) \end{cases} \quad [4.11]$$

where the quantities $\bar{\mathbf{K}}^{(i)}_{A\to q}$, $\bar{\mathbf{K}}^{(i)}_{B\to q}$, $\bar{\mathbf{K}}^{(i)}_{A\to\omega_b}$, $\bar{\mathbf{K}}^{(i)}_{B\to\omega_b}$, $\bar{\mathbf{K}}^{(1)}_{A\to a_s}$, $\bar{\mathbf{K}}^{(1)}_{B\to a_s}$, $\bar{\mathbf{K}}^{(1)}_{A\to b_s}$ and $\bar{\mathbf{K}}^{(1)}_{B\to b_s}$ are arbitrary 1×3 gain matrices that determine the convergence properties of the estimation scheme. In the invariant observer in equation [4.11], the invariant output errors are marked in green, and the basis vectors of the invariant frame are marked in blue.

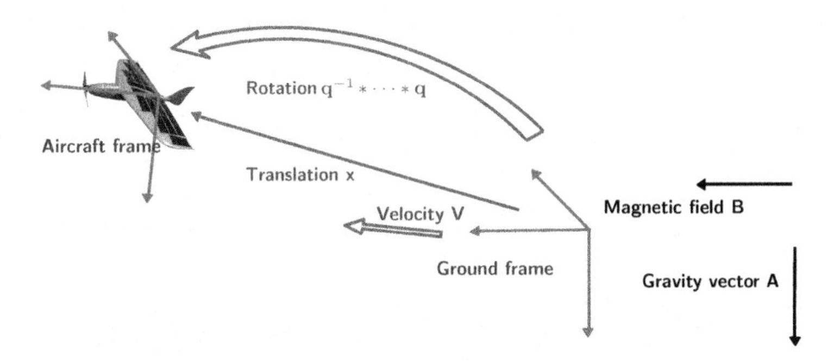

Figure 4.2. *INS – illustration of the invariant Galilean transformations. For a color version of this figure, see www.iste.co.uk/condomines/kalman.zip*

4.4.5. *Invariant filter for the INS*

Our next task is to develop an invariant observer to solve the INS estimation problem. Additional velocity and position information is now

1 For a color version of this equation, see www.iste.co.uk/condomines/kalman.zip.

provided by a GPS and a barometer, supplementing the existing measurements from the accelerometers and magnetometers. By taking advantage of the Galilean invariance properties of the problem, the model equations can be expressed equivalently in both aircraft coordinates and ground coordinates. The INS model equations listed below give a description of the aircraft dynamics and measurements in ground coordinates. As before, the nonlinear state representation associated with the model $\mathcal{M}_{ins}^{+}$ may be compactly written as $\dot{\mathbf{x}} = f(\mathbf{x}, \mathbf{u})$ and $\mathbf{y} = h(\mathbf{x}, \mathbf{u})$, where $\mathbf{x} = (q^T\ V^T\ X^T\ \omega_b^T\ a_s\ h_b)^T$, $\mathbf{u} = (\omega_m^T\ a_m^T)^T$ and $\mathbf{y} = (y_V^T\ y_X^T\ y_h\ y_B^T)^T$ denote the vectors of system states, inputs and outputs, respectively. Recall that:

$$\mathcal{M}_{ins}^{+} \begin{cases} \dot{q} = \frac{1}{2} q * (\omega_m - \omega_b) \\ \dot{V} = A + \frac{1}{a_s} q * a_m * q^{-1} \\ \dot{X} = V & \text{(evolution)} \\ \dot{\omega}_b = 0 \\ \dot{a}_s = 0 \\ \dot{h}_b = 0 \\ \begin{pmatrix} y_V \\ y_X \\ y_h \\ y_B \end{pmatrix} = \begin{pmatrix} V \\ X \\ X_z - h_b \\ q^{-1} * B * q \end{pmatrix} & \text{(observation)} \end{cases} \quad [4.12]$$

As before with the AHRS model, given the model equations listed in equation [4.12] and the Lie group $G = \mathbb{H}_1 \times \mathbb{R}^{11}$ acting on the entire state space of the system dynamics, we can use the following local group transformations to verify that the INS model is also G-invariant and G-equivariant. Thus, $\forall g = (q_0^T\ V_0^T\ X_0^T\ \omega_0^T\ a_0\ h_0)^T \in G$:

$$\varphi_g(\mathbf{x}) = \begin{pmatrix} q * q_0 \\ V + V_0 \\ X + X_0 \\ q_0^{-1} * \omega_b * q_0 + \omega_0 \\ a_s \cdot a_0 \\ h_b + h_0 \end{pmatrix},$$

$$\psi_g(\mathbf{u}) = \begin{pmatrix} q_0^{-1} * \omega_m * q_0 + \omega_0 \\ a_0 \cdot q_0^{-1} * a_m * q_0 \end{pmatrix},$$

$$\rho_g(\mathbf{y}) = \begin{pmatrix} y_V + V_0 \\ y_X + X_0 \\ y_h - h_0 + < X_0, e_3 > \\ q_0^{-1} * y_B * q_0 \end{pmatrix}. \quad [4.13]$$

Each transformation is once again equivalent to either a rotation, a translation or the composition of a rotation and a translation, allowing us to rewrite the vector quantities $\mathbf{x}$, $\mathbf{u}$ and $\mathbf{y}$ in either ground or aircraft coordinates. To find the invariants of the problem, we can apply the method of moving frames to determine the parametrization of the group G that sends the state vector $\mathbf{x}$ to the identity element. In the special case where $g = \mathbf{x}^{-1} = ((q^{-1})^T \ -V^T \ -X^T \ -(q * \omega_b * q^{-1})^T \ a_s^{-1} \ -h_b + < X, e_3 >)^T$, we find:

$$\varphi_{\mathbf{x}^{-1}}(\mathbf{x}) = \begin{pmatrix} q * q^{-1} \\ V - V \\ X - X \\ q * \omega_b * q^{-1} - q * \omega_b * q^{-1} \\ a_s \cdot a_s^{-1} \\ h_b + < X, e_3 > -(h_b + < X, e_3 >) \end{pmatrix} = \begin{pmatrix} 1 \\ \vec{0} \\ \vec{0} \\ \vec{0} \\ 1 \\ 0 \end{pmatrix} = e. [4.14]$$

Solving the normalization equations allows us to define the moving frame γ as $\gamma(\mathbf{x}) = \mathbf{x}^{-1}$, where $\mathbf{x}^{-1}$ is the parameter that yields the invariants. Now, given the standard basis of $\mathbb{R}^3$, $(e_i)_{i \in [\![1;3]\!]}$, as well as the model $\mathcal{M}_{ins}^{+}$, expressions for the vectors that generate the invariant frame may be obtained by evaluating each of the following 14 vector equations at the point $(\hat{q}^T \ \hat{V}^T \ \hat{X}^T \ \hat{\omega}_b^T \ \hat{a}_s \ \hat{h}_b)^T$:

$$\left[D\varphi_{\gamma(\mathbf{x})} \begin{pmatrix} q \\ V \\ X \\ \omega_b \\ a_s \\ h_b \end{pmatrix} \right] .w_j(\mathbf{x}) = \begin{pmatrix} 0 \\ e_i \\ \mathbf{0} \\ \mathbf{0} \\ \mathbf{0} \\ 0 \\ 0 \end{pmatrix}, \begin{pmatrix} 0 \\ \mathbf{0} \\ e_i \\ \mathbf{0} \\ \mathbf{0} \\ 0 \\ 0 \end{pmatrix}, \begin{pmatrix} 0 \\ \mathbf{0} \\ \mathbf{0} \\ e_i \\ \mathbf{0} \\ 0 \\ 0 \end{pmatrix}, \begin{pmatrix} 0 \\ \mathbf{0} \\ \mathbf{0} \\ \mathbf{0} \\ e_i \\ 0 \\ 0 \end{pmatrix}, \begin{pmatrix} 0 \\ \mathbf{0} \\ \mathbf{0} \\ \mathbf{0} \\ \mathbf{0} \\ 1 \\ 0 \end{pmatrix}, \begin{pmatrix} 0 \\ \mathbf{0} \\ \mathbf{0} \\ \mathbf{0} \\ \mathbf{0} \\ 0 \\ 1 \end{pmatrix},$$

$$i \in [\![1;3]\!]. \quad [4.15]$$

As before, after inverting the Jacobian matrix $D\varphi_{\gamma(\mathbf{x})}(\mathbf{x})$ by applying the local inversion theorem to deduce that $[D\varphi_{\gamma(\mathbf{x})}(\mathbf{x})]^{-1} = D\varphi_{\gamma^{-1}(\mathbf{x})}(\mathbf{x})$, we find the following invariant vector field:

$$\begin{pmatrix} w_i^q(\mathbf{x}) \\ w_i^V(\mathbf{x}) \\ w_i^X(\mathbf{x}) \\ w_i^{\omega_b}(\mathbf{x}) \\ w^{a_s}(\mathbf{x}) \\ w^{b_s}(\mathbf{x}) \end{pmatrix} = \begin{pmatrix} 0 \\ e_i * q \\ \hline \mathbf{0} \\ \hline \mathbf{0} \\ \hline \mathbf{0} \\ \hline 0 \\ \hline 0 \end{pmatrix}, \begin{pmatrix} 0 \\ \mathbf{0} \\ \hline e_i \\ \hline \mathbf{0} \\ \hline \mathbf{0} \\ \hline 0 \\ \hline 0 \end{pmatrix}, \begin{pmatrix} 0 \\ \mathbf{0} \\ \hline \mathbf{0} \\ \hline e_i \\ \hline \mathbf{0} \\ \hline 0 \\ \hline 0 \end{pmatrix}, \begin{pmatrix} 0 \\ \mathbf{0} \\ \hline \mathbf{0} \\ \hline \mathbf{0} \\ \hline q^{-1} * e_i * q \\ \hline 0 \\ \hline 0 \end{pmatrix}, \begin{pmatrix} 0 \\ \mathbf{0} \\ \hline \mathbf{0} \\ \hline \mathbf{0} \\ \hline \mathbf{0} \\ \hline a_s \\ \hline 0 \end{pmatrix}, \begin{pmatrix} 0 \\ \mathbf{0} \\ \hline \mathbf{0} \\ \hline \mathbf{0} \\ \hline \mathbf{0} \\ \hline 0 \\ \hline h_b \end{pmatrix},$$

$$i \in [\![1;3]\!]. \qquad [4.16]$$

4.4.6. *Formulating the invariant observer for the INS*

By following the methodology presented above, equation [4.5], which describes a general structure valid for any arbitrary invariant observer, can be refined to the special case of the INS as follows[2]:

$$\begin{cases} \dot{\hat{q}} = \frac{1}{2}\hat{q} * (\omega_m - \hat{\omega}_b) + \\ \qquad \sum_{i=1}^{3}\left(\sum_{j=1}^{3}(\bar{\mathbf{K}}_{V\to q}^{(ij)}E_V + \bar{\mathbf{K}}_{X\to q}^{(ij)}E_X + \bar{\mathbf{K}}_{B\to q}^{(ij)}E_B) + \bar{\mathbf{K}}_{h\to q}^{(ij)}E_h\right) \cdot e_i * \hat{q} \\ \dot{\hat{V}} = A + \frac{\hat{q} * a_m * \hat{q}^{-1}}{\hat{a}_s} + \\ \qquad \sum_{i=1}^{3}\left(\sum_{j=1}^{3}(\bar{\mathbf{K}}_{V\to V}^{(ij)}E_V + \bar{\mathbf{K}}_{X\to V}^{(ij)}E_X + \bar{\mathbf{K}}_{B\to V}^{(ij)}E_B) + \bar{\mathbf{K}}_{h\to V}^{(ij)}E_h\right) \cdot e_i \\ \dot{\hat{X}} = \hat{V} + \sum_{i=1}^{3}\left(\sum_{j=1}^{3}(\bar{\mathbf{K}}_{V\to X}^{(ij)}E_V + \bar{\mathbf{K}}_{X\to X}^{(ij)}E_X + \bar{\mathbf{K}}_{B\to X}^{(ij)}E_B) + \bar{\mathbf{K}}_{h\to X}^{(ij)}E_h\right) \cdot e_i \\ \dot{\hat{\omega}}_b = \hat{q}^{-1} * \sum_{i=1}^{3}\left(\sum_{j=1}^{3}(\bar{\mathbf{K}}_{V\to\omega_b}^{(ij)}E_V + \bar{\mathbf{K}}_{X\to\omega_b}^{(ij)}E_X + \bar{\mathbf{K}}_{B\to\omega_b}^{(ij)}E_B) + \bar{\mathbf{K}}_{h\to\omega_b}^{(ij)}E_h\right) \cdot e_i * \hat{q} \\ \dot{\hat{a}}_s = \hat{a}_s \cdot \left(\sum_{j=1}^{3}(\bar{\mathbf{K}}_{V\to a_s}^{(j)}E_V + \bar{\mathbf{K}}_{X\to a_s}^{(j)}E_X + \bar{\mathbf{K}}_{B\to a_s}^{(j)}E_B) + \bar{\mathbf{K}}_{h\to a_s}^{(j)}E_h\right) \\ \dot{\hat{h}}_b = \sum_{j=1}^{3}(\bar{\mathbf{K}}_{V\to b_s}^{(j)}E_V + \bar{\mathbf{K}}_{X\to b_s}^{(j)}E_X + \bar{\mathbf{K}}_{B\to b_s}^{(j)}E_B) + \bar{\mathbf{K}}_{h\to b_s}^{(j)}E_h \end{cases}$$

2 For a color version of this equation, see www.iste.co.uk/condomines/kalman.zip.

where the invariant output error E is defined by:

$$\begin{pmatrix} \mathsf{E}_V \\ \mathsf{E}_X \\ \mathsf{E}_B \\ \mathsf{E}_h \end{pmatrix} = h\left(e = \varphi_{\hat{\mathbf{x}}^{-1}}(\hat{\mathbf{x}}), \psi_{\hat{\mathbf{x}}^{-1}}(\mathbf{u})\right) - \rho_{\hat{\mathbf{x}}^{-1}}(\mathbf{y})$$

$$= \begin{pmatrix} \hat{V} - y_v \\ \hat{X} - y_x \\ < \hat{X}, e_3 > -\hat{h}_b - y_h \\ B - \hat{q} * y_B * \hat{q}^{-1} \end{pmatrix}. \qquad [4.17]$$

In these filter equations, the components of the invariant frame are marked in blue, and the various gain matrices $\bar{\mathbf{K}}$ are marked in green. The elements of these gain matrices can be chosen arbitrarily to configure the estimator to have certain target dynamics. With our constructive approach, we can derive a set of analytic relations for the invariant observers of the AHRS and INS models that directly (and hence analytically) specify the range of admissible values for the gain terms. This is in fact precisely why the approach is described as constructive. It immediately follows that we can guarantee the convergence of the estimation errors (in terms of the invariants of the estimation problem) while simultaneously being able to freely configure the dynamics of the estimator.

4.5. Invariant state estimation error

The invariance properties of a state estimator, which we would ideally like to preserve the intrinsic symmetries of the system, are closely intertwined with the invariant state estimation error and its dynamics. In fact, it can be shown that the dynamics of the invariant state estimation error depend solely on this error itself, as well as the complete set of fundamental system invariants defined by $I(\hat{\mathbf{x}}, \mathbf{u}) = \psi_{\gamma(\hat{\mathbf{x}})}(\mathbf{u}) = \psi_{\hat{\mathbf{x}}^{-1}}(\mathbf{u})$. Thus, the only dependency of these dynamics on the system trajectories and estimates lies in the nonlinearity of I. This stands in stark contrast to most other conventional nonlinear estimation algorithms such as EKF and UKF, for which the dynamics of the estimation error directly depend on the trajectory of the aircraft. This enlarges the domain of convergence of the invariant observers, ultimately making them more robust estimators.

Thus, given a dynamic system following a nearly constant reference trajectory, i.e. a trajectory for which the fundamental invariants $I(\hat{\mathbf{x}}, \mathbf{u})$ are time independent, $I(\hat{\mathbf{x}}, \mathbf{u}) = c$, it can be shown that the gain matrices $\bar{\mathbf{K}}$ converge to fixed values on any trajectory $\mathbf{x}'$ such that $I(\mathbf{x}', \mathbf{u}) = c$. Furthermore, in the simple case of an AHRS, and given certain decoupling hypotheses on the components of the state vector $\mathbf{x}$, the dynamics of the invariant state estimation error $\eta(\mathbf{x}, \hat{\mathbf{x}})$ can be shown to satisfy a so-called *autonomous* equation, i.e. an equation of the form $\dot{\eta} = \Upsilon(\eta, c)$. Any such filter is characterized by a large domain of convergence for various configurations of the gain values. For the AHRS, the invariant state estimation error vector $\eta(\mathbf{x}, \hat{\mathbf{x}})$ may be defined by the following expression:

$$\begin{aligned}\eta(\mathbf{x}, \hat{\mathbf{x}}) &= \varphi_{\mathbf{x}^{-1}}(\hat{\mathbf{x}}) - \varphi_{\mathbf{x}^{-1}}(\mathbf{x}) \\ &= \begin{pmatrix} \hat{q} * q^{-1} \\ q * \hat{w}_b * q^{-1} - q * w_b * q^{-1} \\ \hat{a}_s / a_s \\ \hat{b}_s / b_s \end{pmatrix} - \begin{pmatrix} 1 \\ \vec{0} \\ 1 \\ 1 \end{pmatrix} \\ &= \begin{pmatrix} \hat{q} * q^{-1} - \mathbf{1} \\ q * (\hat{w}_b - w_b) * q^{-1} \\ \hat{a}_s / a_s - 1 \\ \hat{b}_s / b_s - 1 \end{pmatrix} = \begin{pmatrix} \alpha \\ \beta \\ \mu \\ \nu \end{pmatrix} \sim \mathbf{x}^{-1} \hat{\mathbf{x}}. \end{aligned} \qquad [4.18]$$

The dynamics of this invariant state estimation error may be deduced from equation [4.18] by differentiating with respect to time as follows[3]:

$$\begin{cases} \dot{\alpha} = -\frac{1}{2} \alpha * \beta + \left(\bar{\mathbf{K}}_{A \to q} \cdot \mathsf{E}_A + \bar{\mathbf{K}}_{B \to q} \cdot \mathsf{E}_B \right) * \alpha \\ \dot{\beta} = (\alpha^{-1} * I * \alpha) \times \beta + \alpha^{-1} * \left(\bar{\mathbf{K}}_{A \to \omega_b} \cdot \mathsf{E}_A + \bar{\mathbf{K}}_{B \to \omega_b} \cdot \mathsf{E}_B \right) * \alpha \\ \dot{\mu} = -\mu \cdot (\bar{\mathbf{K}}_{A \to a_s} \cdot \mathsf{E}_A + \bar{\mathbf{K}}_{B \to a_s} \cdot \mathsf{E}_B) \\ \dot{\nu} = -\nu \cdot (\bar{\mathbf{K}}_{A \to b_s} \cdot \mathsf{E}_A + \bar{\mathbf{K}}_{B \to b_s} \cdot \mathsf{E}_B) \end{cases} \qquad [4.19]$$

In order to configure the gain matrices $\bar{\mathbf{K}}$ of the observers $\mathcal{O}_{\mathcal{M}^+_{ahrs}}$ and $\mathcal{O}_{\mathcal{M}^+_{ins}}$, we must first linearize the equations describing the dynamics of the invariant state estimation error (including each part of equation [4.19]). Then, we decompose the correction terms with the goal of decoupling the estimates

3 For a color version of this equation, see www.iste.co.uk/condomines/kalman.zip.

of each system state. For example, for the AHRS, we can decouple the estimates of the longitudinal and lateral components of the model. However, finding the various gain terms can be extremely tedious when constructing an invariant observer. "Manual" fine-tuning is often required, undermining the systematicity of the approach, especially when working with models more complex than those presented in this book. Even for the kinematic models considered here, the configuration process is beginning to become difficult. For the observer $\mathcal{O}_{\mathcal{M}_{ins}^{+}}$, if we wish to compute its dynamics directly and analytically, we must first compute the vector of invariant state estimation errors, which has the following expression:

$$
\begin{aligned}
\eta(\mathbf{x}, \hat{\mathbf{x}}) &= \varphi_{\mathbf{x}^{-1}}(\hat{\mathbf{x}}) - \varphi_{\mathbf{x}^{-1}}(\mathbf{x}) \\
&= \begin{pmatrix} \hat{q} * q^{-1} \\ \hat{V} - V \\ \hat{X} - X \\ q * \hat{w}_b * q^{-1} - q * w_b * q^{-1} \\ \hat{a}_s / a_s \\ \hat{h}_b - h_b \end{pmatrix} - \begin{pmatrix} \mathbf{1} \\ \vec{0} \\ \vec{0} \\ \vec{0} \\ 1 \\ 0 \end{pmatrix} \\
&= \begin{pmatrix} \hat{q} * q^{-1} - \mathbf{1} \\ \hat{V} - V \\ \hat{X} - X \\ q * (\hat{w}_b - w_b) * q^{-1} \\ \hat{a}_s / a_s - 1 \\ \hat{h}_b - h_b \end{pmatrix} = \begin{pmatrix} \alpha \\ \beta \\ \mu \\ \nu \\ \lambda \\ \zeta \end{pmatrix} \sim \mathbf{x}^{-1} \hat{\mathbf{x}}.
\end{aligned}
\qquad [4.20]
$$

The dynamics of the invariant state estimation error $\dot{\eta}(\mathbf{x}, \hat{\mathbf{x}})$ are then governed by the following integro-differential system in this specific case[4]:

$$
\begin{cases}
\overbrace{\alpha e_i \alpha^{-1}}^{\cdot} = -\alpha \nu \alpha^{-1} + \\
\qquad 2\big(\bar{\mathbf{K}}_{V \to q} \mathsf{E}_V + \bar{\mathbf{K}}_{X \to q} \mathsf{E}_X + \bar{\mathbf{K}}_{B \to q} \mathsf{E}_B) + \bar{\mathbf{K}}_{h \to q} \mathsf{E}_h\big) \times \alpha e_i \alpha^{-1} \\
\dot{\beta} = \hat{I}_a - \lambda \alpha^{-1} * \hat{I}_a * \alpha + \big(\bar{\mathbf{K}}_{V \to V} \mathsf{E}_V + \bar{\mathbf{K}}_{X \to V} \mathsf{E}_X + \bar{\mathbf{K}}_{B \to V} \mathsf{E}_B + \bar{\mathbf{K}}_{h \to V} \mathsf{E}_h\big) \\
\dot{\mu} = \beta + \big(\bar{\mathbf{K}}_{V \to X} \mathsf{E}_V + \bar{\mathbf{K}}_{X \to X} \mathsf{E}_X + \bar{\mathbf{K}}_{B \to X} \mathsf{E}_B + \bar{\mathbf{K}}_{h \to X} \mathsf{E}_h\big) \\
\dot{\nu} = (\alpha^{-1} * I_\omega * \alpha) \times \nu + \\
\qquad \alpha^{-1} * \big(\bar{\mathbf{K}}_{V \to \omega_b} \mathsf{E}_V + \bar{\mathbf{K}}_{X \to \omega_b} \mathsf{E}_X + \bar{\mathbf{K}}_{B \to \omega_b} \mathsf{E}_B + \bar{\mathbf{K}}_{h \to \omega_b} \mathsf{E}_h\big) * \alpha \\
\dot{\lambda} = -\lambda \cdot \big(\bar{\mathbf{K}}_{V \to a_s} \mathsf{E}_V + \bar{\mathbf{K}}_{X \to a_s} \mathsf{E}_X + \bar{\mathbf{K}}_{B \to a_s} \mathsf{E}_B + \bar{\mathbf{K}}_{h \to a_s} \mathsf{E}_h\big) \\
\dot{\zeta} = \big(\bar{\mathbf{K}}_{V \to h_b} \mathsf{E}_V + \bar{\mathbf{K}}_{X \to h_b} \mathsf{E}_X + \bar{\mathbf{K}}_{B \to a_s} \mathsf{E}_B + \bar{\mathbf{K}}_{h \to h_b} \mathsf{E}_h\big)
\end{cases}
$$

4 For a color version of this equation, see www.iste.co.uk/condomines/kalman.zip.

To overcome the difficulties associated with "manually" configuring the correction gains, recent literature in this field has proposed a generalized EKF-type algorithm for computing the desired values. This algorithm linearizes the dynamics of the invariant state estimation error, allowing an evolution matrix ($\mathbf{A}(\hat{I})$) and an observation matrix ($\mathbf{C}(\hat{I})$) to be established as a function of the invariants of the estimation problem. The extended Kalman filter equations can then be applied to the linearized system of invariant errors in such a way that the Riccati equation is solved by the pair $(\mathbf{A}(\hat{I}), \mathbf{C}(\hat{I}))$. The values of the gain thus obtained are used to correct the predicted state from equation [4.10]. The mathematical formulation of the approach adopted by the IEKF ultimately reduces to a classical EKF algorithm performed with an invariant state estimation error. IEKF gives a second-order approximation of this error. Despite the many benefits of this technique, it can be relatively complex to implement as a result of the linearization required to identify the matrices $\mathbf{A}(\hat{I})$ and $\mathbf{C}(\hat{I})$. Faced with this obstacle, our research led us to develop two distinct methods for estimating the state of dynamic systems in nonlinear settings. Like IEKF, we sought to equip our estimators with the same invariance properties as the observed system. However, to avoid resorting to certain computations judged *a priori* too complex to be embedded and integrated into the Paparazzi project (such as linearizing the model at any given moment in time), our work focused on applying the principles of unscented Kalman filtering in combination with ideas from invariant observer theory. The first method, presented in the next section, was baptized as the IUKF algorithm, and the second was named π-IUKF. The symbol π refers to the so-called *compatibility condition* π that must be introduced to unite the theoretical principles of constructing arbitrary invariant observers with the UKF algorithm. Note that our research also allowed us to formulate factorized variants of the IUKF and π-IUKF algorithms.

4.6. The SR-UKF algorithm

The SR-UKF algorithm provides the basis of the methodological work presented in this book. Below, we give a pseudocode description of this algorithm, with more details about each step in the main loop:

SR-UKF algorithm for estimating the state with additive noise terms
Description of the process: $\mathbf{x}_{k+1} = f(\mathbf{x}_k, \mathbf{u}_k) + \mathbf{w}_k$ Model of the sensors: $\mathbf{y}_k = h(\mathbf{x}_k, \mathbf{u}_k) + \mathbf{v}_k$
► Compute the weights $W_m^{(j)}$, $W_c^{(j)}$, $j \in [\![0 ; 2n]\!]$ (see Scaled UKF)
► Initialize $\hat{\mathbf{x}}_0 = \mathsf{E}[\mathbf{x}_0]$ et $\hat{\mathbf{S}}_0 = \mathsf{chol}(\mathsf{E}[(\mathbf{x}_0 - \hat{\mathbf{x}}_0)(\mathbf{x}_0 - \hat{\mathbf{x}}_0)^T])$
► Compute the $(2n+1)$ sigma points: $\boldsymbol{\mathcal{X}}_{k\|k}^{(0)} = \hat{\mathbf{x}}_{k\|k}, \boldsymbol{\mathcal{X}}_{k\|k}^{(1)}, \ldots, \boldsymbol{\mathcal{X}}_{k\|k}^{(2n)}$ ► Prediction step: ❶ $\forall i \in [\![0 ; 2n]\!]$, $\boldsymbol{\mathcal{X}}_{k+1\|k}^{(i)} = f(\boldsymbol{\mathcal{X}}_{k\|k}^{(i)}, \mathbf{u}_k) \Rightarrow \hat{\mathbf{x}}_{k+1\|k} = \sum_{i=0}^{2n} W_{(m)}^{(i)} \boldsymbol{\mathcal{X}}_{k+1\|k}^{(i)}$ ❷ Find the QR factorization of the predicted covariance of the invariant state errors: $\hat{\mathbf{S}}_{\mathbf{xx},k+1\|k} = \begin{cases} ① \ \mathsf{qr}\left[\sqrt{W_{(c)}^{(1)}} \left((\boldsymbol{\mathcal{X}}_{k+1\|k}^{(1)} - \hat{\mathbf{x}}_{k+1\|k}) \ \cdots \ (\boldsymbol{\mathcal{X}}_{k+1\|k}^{(2n)} - \hat{\mathbf{x}}_{k+1\|k})\right) \ \mathbf{Q}^{1/2}\right] \\ ② \ \mathsf{cholupdate}\left(\hat{\mathbf{S}}_{\mathbf{xx},k+1\|k}, (\boldsymbol{\mathcal{X}}_{k+1\|k}^{(0)} - \hat{\mathbf{x}}_{k+1\|k}), W_{(c)}^{(0)}\right) \end{cases}$ ❸ $\forall i \in [\![0 ; 2n]\!]$, $\hat{\mathbf{y}}_{k+1\|k}^{(i)} = h(\boldsymbol{\mathcal{X}}_{k+1\|k}^{(i)}, \mathbf{u}_k) \Rightarrow \hat{\mathbf{y}}_{k+1\|k} = \sum_{i=0}^{2n} W_{(m)}^{(i)} \hat{\mathbf{y}}_{k+1\|k}^{(i)}$ ❹ Find the QR factorization of the predicted covariance of the invariant output errors: $\hat{\mathbf{S}}_{\mathbf{yy},k+1\|k} = \begin{cases} ① \ \mathsf{qr}\left[\sqrt{W_{(c)}^{(1)}} \left((\hat{\mathbf{y}}_{k+1\|k}^{(1)} - \hat{\mathbf{y}}_{k+1\|k}) \ \cdots \ (\hat{\mathbf{y}}_{k+1\|k}^{(2n)} - \hat{\mathbf{y}}_{k+1\|k})\right) \ \mathbf{R}^{1/2}\right] \\ ② \ \mathsf{cholupdate}\left(\hat{\mathbf{S}}_{\mathbf{yy},k+1\|k}, (\hat{\mathbf{y}}_{k+1\|k}^{(0)} - \hat{\mathbf{y}}_{k+1\|k}), W_{(c)}^{(0)}\right) \end{cases}$ ❺ Predict the cross-covariance of the invariant estimation errors: $\hat{\mathbf{P}}_{\mathbf{xy},k+1\|k} = \sum_{i=0}^{2n} W_{(c)}^{(i)} (\boldsymbol{\mathcal{X}}_{k+1\|k}^{(i)} - \hat{\mathbf{x}}_{k+1\|k})(\hat{\mathbf{y}}_{k+1\|k}^{(i)} - \hat{\mathbf{y}}_{k+1\|k})$ ► Correction/filtering step: ❶ Compute the gain terms: $\mathbf{K}_{k+1} = (\hat{\mathbf{P}}_{\mathbf{xy},k+1\|k} / \hat{\mathbf{S}}_{\mathbf{yy},k+1\|k}^T) / \hat{\mathbf{S}}_{\mathbf{yy},k+1\|k}$ ❷ Compute the new estimated state: $\hat{\mathbf{x}}_{k+1\|k+1} = \hat{\mathbf{x}}_{k+1\|k} + \mathbf{K}_{k+1} \cdot (\mathbf{z}_{k+1} - \hat{\mathbf{y}}_{k+1\|k})$ ❸ Update the matrix $\hat{\mathbf{S}}_{\mathbf{xx}}$: $\hat{\mathbf{S}}_{\mathbf{xx},k+1\|k+1} = \mathsf{cholupdate}\left(\hat{\mathbf{S}}_{\mathbf{xx},k+1\|k+1}, \mathbf{K}_{k+1}\hat{\mathbf{S}}_{\mathbf{yy},k+1\|k}, -1\right)$

SR-UKF offers an alternative to the standard UKF algorithm that is numerically extremely valuable. The most computationally expensive step in the implementation of UKF is the process of finding the sigma points at each moment in time, since the algorithm needs to find the square root matrix $\hat{\mathbf{S}}_{k|k}$ of the estimated covariance of the state estimation error $\hat{\mathbf{P}}_{k|k}$, which satisfies

$\hat{\mathbf{P}}_{k|k} = \hat{\mathbf{S}}_{k|k}\hat{\mathbf{S}}_{k|k}^T$. This can be done by Cholesky factorization. However, to avoid systematically invoking linear algebra, the idea of SR-UKF is to update the square root matrix directly instead of the covariance matrices. As a result, SR-UKF only requires one single Cholesky factorization, performed on the initial matrix $\hat{\mathbf{P}}_0$. Knowing $\hat{\mathbf{S}}_{k|k}$ at each moment in time makes the sigma points significantly easier to compute. Moreover, to avoid needing to calculate multiple matrix products and their Cholesky factorizations during the prediction and correction steps, SR-UKF takes advantage of three other very powerful numerical techniques from linear algebra: ❶ QR factorization (qr operator); ❷ rank-one updating of the Cholesky factor (cholupdate operator); ❸ and pseudo-inversion by least squares (/ operator).

4.7. First reformulation of unscented Kalman filtering in an invariant setting: the IUKF algorithm

4.7.1. *Foundational principles*

If the dynamics of the observed system have invariance properties (symmetries), we cannot simply construct an estimator of the system state with analogous properties directly from the basic equations of the SR-UKF algorithm. For convergence, it would be extremely desirable for any candidate estimator filters to satisfy the same invariance properties as the system itself, in the same spirit as the invariant observers of the IEKF algorithm. To achieve this, we need to modify the algorithm listed above. Our objective is to adapt the SR-UKF algorithm so that it yields an invariant estimator. From the same principles and computation steps as the SR-UKF algorithm (see section [4.6]), a "natural" reformulation of the equations aiming to adapt the method for estimation in an invariant setting can be obtained simply by redefining the error terms used in the standard algorithm, which are presented in the previous section. The linear error terms $\hat{\mathbf{x}} - \mathbf{x}$ and $\mathbf{z} - \hat{\mathbf{y}}$ conventionally used for Kalman filtering do not preserve any of the symmetries and invariance properties of the system. Instead, we can consider the following invariant state and output error terms:

$$\begin{cases} \hat{\mathbf{x}} - \mathbf{x} \longrightarrow \eta(\mathbf{x}, \hat{\mathbf{x}}) = \varphi_{\gamma(\mathbf{x})}(\mathbf{x}) - \varphi_{\gamma(\mathbf{x})}(\hat{\mathbf{x}}) \\ \mathbf{z} - \hat{\mathbf{y}} \longrightarrow \mathsf{E}(\mathbf{z}, \hat{\mathbf{x}}, \hat{\mathbf{y}}) = \rho_{\gamma(\hat{\mathbf{x}})}(\hat{\mathbf{y}}) - \rho_{\gamma(\hat{\mathbf{x}})}(\mathbf{z}) \end{cases} \qquad [4.21]$$

In equation [4.21], $\mathbf{x}$ is the "true" state of the system and $\hat{\mathbf{x}}$ is the estimate of this state. The terms $\hat{\mathbf{y}}$ and $\mathbf{z}$, respectively, denote the estimated output and the vector of noisy measurements.

REMARK 4.1.– It is easy to check that these errors are invariant. Indeed, $\forall g \in G$, we have that:

$$\begin{aligned}\eta(\varphi_g(\mathbf{x}), \varphi_g(\hat{\mathbf{x}})) &= \varphi_{\gamma(\varphi_g(\mathbf{x}))}(\varphi_g(\mathbf{x})) - \varphi_{\gamma(\varphi_g(\mathbf{x}))}(\varphi_g(\hat{\mathbf{x}}))\\ &= \varphi_{\gamma(\varphi_g(\mathbf{x}))g}(\mathbf{x}) - \varphi_{\gamma(\varphi_g(\mathbf{x}))g}(\hat{\mathbf{x}})\\ &= \varphi_{\gamma(\mathbf{x})}(\mathbf{x}) - \varphi_{\gamma(\mathbf{x})}(\hat{\mathbf{x}})\\ &= \eta(\mathbf{x}, \hat{\mathbf{x}}) \qquad \square\end{aligned}$$

and:

$$\begin{aligned}\mathsf{E}(\rho_g(\mathbf{z}), \varphi_g(\hat{\mathbf{x}}), \rho_g(\hat{\mathbf{y}})) &= \rho_{\gamma(\varphi_g(\hat{\mathbf{x}}))}(\rho_g(\hat{\mathbf{y}})) - \rho_{\gamma(\varphi_g(\hat{\mathbf{x}}))}(\rho_g(\mathbf{z}))\\ &= \rho_{\gamma(\varphi_g(\hat{\mathbf{x}}))g}(\hat{\mathbf{y}}) - \rho_{\gamma(\varphi_g(\hat{\mathbf{x}}))g}(\mathbf{z})\\ &= \rho_{\gamma(\hat{\mathbf{x}})}(\hat{\mathbf{y}}) - \varphi_{\gamma(\hat{\mathbf{x}})}(\mathbf{z})\\ &= \mathsf{E}(\mathbf{z}, \hat{\mathbf{x}}, \hat{\mathbf{y}}) \qquad \square\end{aligned}$$

Whenever the group action is full-rank and transitive (i.e. $\mathsf{dim}(G) = \mathsf{dim}(\mathcal{X}) = n$), we saw earlier that G can be identified with the state space $\mathcal{X} = \mathbb{R}^n$ in such a way that the local transformation φ_g is viewed as the left-multiplication mapping $\varphi_g(\mathbf{x}) = g \cdot \mathbf{x}$. Solving the normalization equations to obtain $\varphi_g(\mathbf{x}) = g \cdot \mathbf{x} = e$, where e is the identity element of the group G, gives us the moving frame $\gamma(\mathbf{x}) = \mathbf{x}^{-1}$ as a solution. The errors in equation [4.21] can now be written as:

$$\begin{cases}\eta(\mathbf{x}, \hat{\mathbf{x}}) = \varphi_{\mathbf{x}^{-1}}(\mathbf{x}) - \varphi_{\mathbf{x}^{-1}}(\hat{\mathbf{x}}) = e - \mathbf{x}^{-1} \cdot \hat{\mathbf{x}}\\ \mathsf{E}(\mathbf{z}, \hat{\mathbf{x}}, \hat{\mathbf{y}}) = \rho_{\hat{\mathbf{x}}^{-1}}(\hat{\mathbf{y}}) - \rho_{\hat{\mathbf{x}}^{-1}}(\mathbf{z})\end{cases} \qquad [4.22]$$

Illustration for the AHRS: Simply by considering the evolution equations of the model $\mathcal{M}_{ahrs}$, the group transformation $\varphi_{\mathbf{x}_0 \in \mathcal{X}}(\mathbf{x})$, and the inverse $\mathbf{x}^{-1}$, defined as follows:

$$\begin{cases}\dot{q} = \frac{1}{2} q * (w_m - w_b)\\ \dot{w}_b = 0\\ \dot{a}_s = 0\\ \dot{b}_s = 0\end{cases}, \varphi_{\mathbf{x}_0 \in \mathcal{X}}(\mathbf{x}) = \begin{pmatrix} q * q_0\\ q_0^{-1} * w_b * q_0 + w_0\\ a_0 a_s\\ b_0 b_s\end{pmatrix},$$

$$\mathbf{x}^{-1} = \begin{pmatrix} q^{-1} \\ -q * w_b * q^{-1} \\ 1/a_s \\ 1/b_s \end{pmatrix},$$

we can already deduce the analytic expression of $\eta(\mathbf{x}, \hat{\mathbf{x}})$:

$$\begin{aligned} \eta(\mathbf{x}, \hat{\mathbf{x}}) &= e - \varphi_{\mathbf{x}^{-1}}(\hat{\mathbf{x}}) = \begin{pmatrix} \mathbf{1} \\ 0 \\ 1 \\ 1 \end{pmatrix} - \begin{pmatrix} \hat{q} * q^{-1} \\ q * (\hat{w}_b - w_b) * q^{-1} \\ \hat{a}_s/a_s \\ \hat{b}_s/b_s \end{pmatrix} \\ &= \begin{pmatrix} \mathbf{1} - \hat{q} * q^{-1} \\ q * (w_b - \hat{w}_b) * q^{-1} \\ 1 - \hat{a}_s/a_s \\ 1 - \hat{b}_s/b_s \end{pmatrix}. \end{aligned} \quad [4.23]$$

In equation [4.23], $\mathbf{1}=(1\ 0\ 0\ 0)^T$ denotes the quaternion of norm one that acts as the identity element under multiplication $(*)$.

Illustration for the INS: Similarly, by considering

$$\begin{cases} \dot{q} = \frac{1}{2} q * (w_m - w_b) \\ \dot{V} = A + \frac{1}{a_s} q * a_m * q^{-1} \\ \dot{X} = V \\ \dot{w}_b = 0 \\ \dot{a}_s = 0 \\ \dot{h}_b = 0 \end{cases}, \varphi_{\mathbf{x}_0 \in \mathcal{X}}(\mathbf{x}) = \begin{pmatrix} q * q_0 \\ V + V_0 \\ X + X_0 \\ q_0^{-1} * w_b * q_0 + w_0 \\ a_0 a_s \\ h_b + h_0 \end{pmatrix}, \mathbf{x}^{-1}$$

$$= \begin{pmatrix} q^{-1} \\ -V \\ -X \\ -q * w_b * q^{-1} \\ 1/a_s \\ -h_b \end{pmatrix},$$

we obtain:

$$\eta(\mathbf{x}, \hat{\mathbf{x}}) = e - \varphi_{\mathbf{x}^{-1}}(\hat{\mathbf{x}}) = \begin{pmatrix} \mathbf{1} \\ 0 \\ 0 \\ 0 \\ 1 \\ 0 \end{pmatrix} - \begin{pmatrix} \hat{q} * q^{-1} \\ \hat{V} - V \\ \hat{X} - X \\ q * (\hat{w}_b - w_b) * q^{-1} \\ \hat{a}_s / a_s \\ \hat{h}_b - h_b \end{pmatrix}$$
$$= \begin{pmatrix} \mathbf{1} - \hat{q} * q^{-1} \\ V - \hat{V} \\ X - \hat{X} \\ q * (w_b - \hat{w}_b) * q^{-1} \\ 1 - \hat{a}_s / a_s \\ h_b - \hat{h}_b \end{pmatrix}. \quad [4.24]$$

We are now in a position to modify our definition of the covariance matrix of the estimation errors, which the SR-UKF approximated by constructing the $(2n+1)$ sigma points $\boldsymbol{\mathcal{X}}_{k|k}^{(i)}$, $i \in [\![1 \,; (2n+1)]\!]$ at each time step $k \in \mathbb{Z}^*$, so that we can instead apply it to the new errors defined in equation [4.22]. Below, the modified version is called the covariance matrix of the invariant estimation errors. Accordingly, the prediction step of the algorithm presented in section 4.21 now becomes :

---

(1) **Prediction step** $(\forall k \geq 1)$

– Apply the mapping f to the $(2n+1)$ sigma points:

$$\boldsymbol{\mathcal{X}}_{k+1|k} = [\boldsymbol{\mathcal{X}}_{k+1|k}^{(0)} \ \boldsymbol{\mathcal{X}}_{k+1|k}^{(1)} \ \cdots \ \boldsymbol{\mathcal{X}}_{k+1|k}^{(2n)}] = f(\boldsymbol{\mathcal{X}}_{k|k}, \mathbf{u}_k)$$

– Compute the predicted state:

$$\hat{\mathbf{x}}_{k+1|k} = \sum_{i=0}^{2n} W_{(m)}^{(i)} \boldsymbol{\mathcal{X}}_{k+1|k}^{(i)}$$

• Find the QR factorization of the covariance matrix of the invariant state estimation error as follows:

$$\begin{cases} \mathbf{S}_{k+1|k} = \\ \text{qr}\left(\left[\sqrt{W_{(c)}^{(1)}}\left(\eta(\boldsymbol{\mathcal{X}}_{k+1|k}^{(1)}, \hat{\mathbf{x}}_{k+1|k}) \ldots \eta(\boldsymbol{\mathcal{X}}_{k+1|k}^{(2n)}, \hat{\mathbf{x}}_{k+1|k})\right) \sqrt{\mathbf{W}_k^{(1)}}\right]\right) \\ \mathbf{S}_{k+1|k} = \text{cholupdate}\left(\mathbf{S}_{k+1|k}, \eta(\boldsymbol{\mathcal{X}}_{k+1|k}^{(0)}, \hat{\mathbf{x}}_{k+1|k}), W_{(c)}^{(0)}\right) \end{cases}$$

where,
$\forall i \in [\![0\,;2n]\!]$, $\eta(\boldsymbol{\mathcal{X}}_{k+1|k}^{(i)}, \hat{\mathbf{x}}_{k+1|k}) = \varphi_{\boldsymbol{\mathcal{X}}_{k+1|k}^{(i)-1}}(\boldsymbol{\mathcal{X}}_{k+1|k}^{(i)}) - \varphi_{\boldsymbol{\mathcal{X}}_{k+1|k}^{(i)-1}}(\hat{\mathbf{x}}_{k+1|k}) = e - \boldsymbol{\mathcal{X}}_{k+1|k}^{(i)-1} \cdot \hat{\mathbf{x}}_{k+1|k}$

– Apply the mapping h to the $(2n+1)$ sigma points in the matrix $\boldsymbol{\mathcal{X}}_{k+1|k}$:

$$\hat{\mathbf{Y}}_{k+1|k} = [\hat{\mathbf{y}}_{k+1|k}^{(0)}\, \hat{\mathbf{y}}_{k+1|k}^{(1)} \cdots \hat{\mathbf{y}}_{k+1|k}^{(2n)}] = h(\boldsymbol{\mathcal{X}}_{k+1|k}, \mathbf{u}_k)$$

– Compute the predicted output:

$$\hat{\mathbf{y}}_{k+1|k} = \sum_{i=0}^{2n} W_{(m)}^{(i)} \hat{\mathbf{y}}_{k+1|k}^{(i)}$$

The factorization step of the covariance matrix of the state estimation error leads to a set of $(2n+1)$ invariant errors defined between each sigma point and the predicted state. Thus, the covariance has now been conceptually redefined in terms of the invariant state error instead of a linear error term. This is the most important nuance compared to classical unscented Kalman filtering. Other than this key modification, the prediction step is essentially identical to the original algorithm. Next, we need to modify the correction step accordingly to compute the gain terms:

(2) **Correction step** $(\forall k \geq 1)$

• Find the QR factorization of the covariance matrix of the invariant output estimation errors:

$$\begin{cases} \mathbf{S}_{\mathbf{y},k+1|k} = \\ \mathsf{qr}\left(\left[\sqrt{W_{(c)}^{(1)}}\left(\mathsf{E}(\hat{\mathbf{y}}_{k+1|k}, \boldsymbol{\mathcal{X}}_{k+1|k}^{(1)}, \hat{\mathbf{y}}_{k+1|k}^{(1)})\right.\right.\right. \\ \qquad \left.\left.\left. \ldots \mathsf{E}(\hat{\mathbf{y}}_{k+1|k}, \boldsymbol{\mathcal{X}}_{k+1|k}^{(2n)}, \hat{\mathbf{y}}_{k+1|k}^{(2n)})\right) \ \sqrt{\mathbf{V}_{k+1}^{(1)}}\right]\right) \\ \mathbf{S}_{\mathbf{y},k+1|k} = \mathsf{cholupdate}\left(\mathbf{S}_{\mathbf{y},k+1|k}, \mathsf{E}(\hat{\mathbf{y}}_{k+1|k}, \boldsymbol{\mathcal{X}}_{k+1|k}^{(0)}, \hat{\mathbf{y}}_{k+1|k}^{(0)}), W_{(c)}^{(0)}\right) \end{cases}$$

where, $\forall i \in [\![0 ; 2n]\!]$,

$$\begin{aligned} \mathsf{E}(\hat{\mathbf{y}}_{k+1|k}, \boldsymbol{\mathcal{X}}_{k+1|k}^{(i)}, \hat{\mathbf{y}}_{k+1|k}^{(i)}) &= \rho_{\boldsymbol{\mathcal{X}}_{k+1|k}^{(i)-1}}(\hat{\mathbf{y}}_{k+1|k}^{(i)}) - \rho_{\boldsymbol{\mathcal{X}}_{k+1|k}^{(i)-1}}(\hat{\mathbf{y}}_{k+1|k}) \\ &= \rho_{\boldsymbol{\mathcal{X}}_{k+1|k}^{(i)-1}}(h(\boldsymbol{\mathcal{X}}_{k+1|k}^{(i)}, \mathbf{u}_k)) - \rho_{\boldsymbol{\mathcal{X}}_{k+1|k}^{(i)-1}}(\hat{\mathbf{y}}_{k+1|k}) \\ &= h(\varphi_{\boldsymbol{\mathcal{X}}_{k+1|k}^{(i)-1}}(\boldsymbol{\mathcal{X}}_{k+1|k}^{(i)}), \psi_{\boldsymbol{\mathcal{X}}_{k+1|k}^{(i)-1}}(\mathbf{u}_k)) \\ &\quad -\rho_{\boldsymbol{\mathcal{X}}_{k+1|k}^{(i)-1}}(\hat{\mathbf{y}}_{k+1|k}) \\ &= h(e, I(\boldsymbol{\mathcal{X}}_{k+1|k}^{(i)}, \mathbf{u}_k)) - \rho_{\boldsymbol{\mathcal{X}}_{k+1|k}^{(i)-1}}(\hat{\mathbf{y}}_{k+1|k}) \end{aligned}$$

• Approximate the cross-covariance matrix between the invariant state and output estimation errors:

$$\begin{aligned} \mathbf{P}_{\mathbf{xy},k+1|k} &= \sum_{i=0}^{2n} W_{(c)}^{(i)} \eta(\boldsymbol{\mathcal{X}}_{k+1|k}^{(i)}, \hat{\mathbf{x}}_{k+1|k}) \mathsf{E}^T(\hat{\mathbf{y}}_{k+1|k}, \boldsymbol{\mathcal{X}}_{k+1|k}^{(i)}, \hat{\mathbf{y}}_{k+1|k}^{(i)}) \\ &= \sum_{i=0}^{2n} W_{(c)}^{(i)} \left(e - \boldsymbol{\mathcal{X}}_{k+1|k}^{(i)-1} \cdot \hat{\mathbf{x}}_{k+1|k}\right) \left(h(e, I(\boldsymbol{\mathcal{X}}_{k+1|k}^{(i)}, \mathbf{u}_k))\right. \\ &\quad \left. -\rho_{\boldsymbol{\mathcal{X}}_{k+1|k}^{(i)-1}}(\hat{\mathbf{y}}_{k+1|k})\right)^T \end{aligned}$$

– Compute the matrix of correction gains:

$$\mathbf{K}_{k+1} = (\mathbf{P}_{\mathbf{xy},k+1|k} / \mathbf{S}_{\mathbf{y},k+1|k}^T) / \mathbf{S}_{\mathbf{y},k+1|k} \in \mathcal{M}_{n \times p}(\mathbb{R})$$

• Compute the estimated state at time $(k+1)$ given the measurement $\mathbf{z}_{k+1}$:

$$\begin{aligned}\hat{\mathbf{x}}_{k+1|k+1} &= \hat{\mathbf{x}}_{k+1|k} + \sum_{i=1}^{n} \mathbf{K}_{k+1}^{(i)} \cdot \mathsf{E}(\hat{\mathbf{y}}_{k+1|k}, \hat{\mathbf{x}}_{k+1|k}, \mathbf{z}_{k+1}) \cdot w(\hat{\mathbf{x}}_{k+1|k}) \\ &= \hat{\mathbf{x}}_{k+1|k} + \sum_{i=1}^{n} \mathbf{K}_{k+1}^{(i)} \cdot \left(\rho_{\hat{\mathbf{x}}_{k+1|k}^{-1}}(\mathbf{z}_{k+1}) - \rho_{\hat{\mathbf{x}}_{k+1|k}^{-1}}(\hat{\mathbf{y}}_{k+1|k})\right) \\ &\cdot w_i(\hat{\mathbf{x}}_{k+1|k})\end{aligned}$$

where, $\forall i \in [\![1;n]\!]$, $\forall \mathbf{x} \in \mathcal{X}$, $w_i(\mathbf{x}) = \left(D\varphi_{\gamma(\mathbf{x})}(\mathbf{x})\right)^{-1} \cdot \partial()/\partial x_i$. The notation $\partial()/\partial x_i$ denotes the ith basis vector in the standard basis of $\mathbb{R}^n$.

– Update the covariance matrix of the invariant state estimation error:

$$\mathbf{S}_{k+1|k+1} = \mathsf{cholupdate}\left(\mathbf{S}_{k+1|k}, \mathbf{K}_{k+1}\mathbf{S}_{\mathbf{y},k+1|k}, -1\right)$$

Thus, the correction step of the standard SR-UKF algorithm requires more extensive modifications. With our new definition of the covariance matrix, which is now associated with the invariant estimation errors, we need to make the following changes:

1) the calculation of the predicted covariance of the output $\mathbf{y}$ now becomes:

$$\mathbf{P}_{\mathbf{y},k+1|k} \propto \mathsf{E}(\hat{\mathbf{y}}_{k+1|k}, \boldsymbol{\mathcal{X}}_{k+1|k}, \hat{\mathbf{Y}}_{k+1|k}),$$

2) and the calculation of the predicted cross-covariance between the estimation errors in the state $\mathbf{x}$ and the output $\mathbf{y}$ is now:

$$\mathbf{P}_{\mathbf{xy},k+1|k} \propto \eta(\boldsymbol{\mathcal{X}}_{k+1|k}, \hat{\mathbf{x}}_{k+1|k}), \mathsf{E}(\hat{\mathbf{y}}_{k+1|k}, \boldsymbol{\mathcal{X}}_{k+1|k}, \hat{\mathbf{Y}}_{k+1|k}).$$

By transitivity, it naturally follows that any correction gains for the state estimate computed using a UKF-type technique are functions of the fundamental invariants of the system $(\psi_{\mathbf{x}^{-1}}(u))$ and the invariant output errors $(\mathsf{E}(\hat{\mathbf{y}}, \mathbf{x}, \cdot))$. The invariant setting defined for the system also requires us to modify the correction equations for the predicted state. The additive correction term now includes: ▶ a gain term that depends on the invariants of the estimation problem; ▶ an invariant innovation term. This correction term is projected onto the invariant frame in such a way that the predicted state can be corrected component by component, i.e. along each of the n vectors in the standard basis of $\mathbb{R}^n$ formed by the invariant vector field $\mathcal{B}(\hat{\mathbf{x}}_{k+1|k}) = \{\omega_i(\hat{\mathbf{x}}_{k+1|k})\}_{i\in[\![1\,;n]\!]}$. This is the most natural approach to adapting the SR-UKF algorithm in order to construct an invariant estimator such that:

– the correction terms are computed by a UKF-type scheme adapted to the invariant setting of the estimation problem that samples the state space using a classical SUT (Scaled Unscented Transform) technique, like conventional sigma-points Kalman filtering;

– the correction terms also preserve the specific symmetries of the system, since they are constructed from an invariant innovation term and gain terms that are functions of the fundamental invariants and the invariant output error.

This first formulation was largely inspired by past research into IEKF [BON 09b].

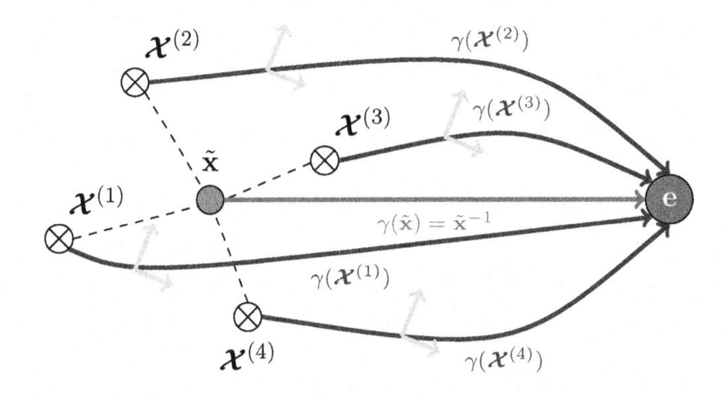

Figure 4.3. *Network of $(2n+1)$ invariant frames associated with the sigma points. For a color version of this figure, see www.iste.co.uk/condomines/kalman.zip*

At this point, we should note that the algorithm presented above uses a multiple parametrization of the transformation group obtained by successively defining the inverse of each sigma point as a parameter of the composite mapping $\phi_g = (\varphi_g, \psi_g, \rho_g)$. This is ultimately equivalent to defining a set of $(2n+1)$ n-dimensional moving frames in the state space that send each sigma point to the identity element e via the local mapping φ_g. Recall that this local mapping may be viewed as a full-rank and transitive action $\varphi_g(\mathbf{x}) = g \cdot \mathbf{x}$ (see Figure 4.3). The algorithm proposed here is generic in the sense that it does not assume any specific form for the equations of the observation model nor the relations which define the group transformation ρ_g. Nevertheless, it can sometimes be useful to extend and specialize the computations in each of the steps listed above in order to make them more explicit. Below, we illustrate this for the special case of the AHRS defined by the model and group transformations from sections 4.4.1 and 4.4.2.

The newly proposed reformulation outlined earlier defines a finite set of $(2n+1)$ invariant state estimation errors $\{\eta(\boldsymbol{\mathcal{X}}^{(i)}_{k+1|k}, \hat{\mathbf{x}}_{k+1|k}) / i \in [\![0\,; 2n]\!]\}$. In the simple case of an AHRS, these errors can be explicitly expressed as follows (see results from section 4.5, up to choice of sign):

$$\forall i \in [\![0\,; 2n]\!],\ \eta(\boldsymbol{\mathcal{X}}^{(i)}, \hat{\mathbf{x}}_{k+1|k}) = \begin{pmatrix} \mathbf{1} - \hat{q}_{k+1|k} * q^{-1}_{\boldsymbol{\mathcal{X}}^{(i)}} \\ q_{\boldsymbol{\mathcal{X}}^{(i)}} * (w_{b,\boldsymbol{\mathcal{X}}^{(i)}} - \hat{w}_{b,k+1|k}) * q^{-1}_{\boldsymbol{\mathcal{X}}^{(i)}} \\ 1 - \hat{a}_{s,k+1|k} / a_{s,\boldsymbol{\mathcal{X}}^{(i)}} \\ 1 - \hat{b}_{s,k+1|k} / b_{s,\boldsymbol{\mathcal{X}}^{(i)}} \end{pmatrix}. \quad [4.25]$$

In equation [4.25], the index $k+1|k$ was deliberately omitted for each of the $(2n+1)$ sigma points to simplify the notation. The same convention is followed below in some of the equations below.

Similarly, the new algorithm introduces a complete set of $(2n+1)$ fundamental invariants which, in the case of the AHRS, satisfy:

$$\begin{aligned} \forall i \in [\![0\,; 2n]\!],\ I(\boldsymbol{\mathcal{X}}^{(i)}_{k+1|k}, \mathbf{u}_k) &= \psi_{\boldsymbol{\mathcal{X}}^{(i)-1}_{k+1|k}}(w_m(k)) \\ &= q_{\boldsymbol{\mathcal{X}}^{(i)}} * w_m(k) * q^{-1}_{\boldsymbol{\mathcal{X}}^{(i)}} \\ &\quad - q_{\boldsymbol{\mathcal{X}}^{(i)}} * w_{b,\boldsymbol{\mathcal{X}}^{(i)}} * q^{-1}_{\boldsymbol{\mathcal{X}}^{(i)}} \\ &= q_{\boldsymbol{\mathcal{X}}^{(i)}} * (w_m(k) - w_{b,\boldsymbol{\mathcal{X}}^{(i)}}) * q^{-1}_{\boldsymbol{\mathcal{X}}^{(i)}}. \end{aligned} \quad [4.26]$$

For the AHRS, there is a straightforward way to express the $(2n+1)$ invariant output errors in terms of the constant vector $(A^T\ B^T)^T$, the transformation group ρ_g and a set of invariant state estimation errors satisfying the following relation for all $\in [\![0\,;2n]\!]$:

$$\mathsf{E}(\hat{\mathbf{y}}, \boldsymbol{\mathcal{X}}^{(i)}, \hat{\mathbf{y}}^{(i)}) = h(e, \mathbf{I}(\boldsymbol{\mathcal{X}}^{(i)}, \mathbf{u}_k)) - \rho_{\boldsymbol{\mathcal{X}}^{(i)-1}}(\hat{\mathbf{y}}_{k+1|k})$$

$$= \begin{pmatrix} A \\ B \end{pmatrix} - \begin{pmatrix} a^{-1}_{s,\boldsymbol{\mathcal{X}}^{(i)}} q_{\boldsymbol{\mathcal{X}}^{(i)}} * \hat{\mathbf{y}}_{A,k+1|k} * q^{-1}_{\boldsymbol{\mathcal{X}}^{(i)}} \\ b^{-1}_{s,\boldsymbol{\mathcal{X}}^{(i)}} q_{\boldsymbol{\mathcal{X}}^{(i)}} * \hat{\mathbf{y}}_{B,k+1|k} * q^{-1}_{\boldsymbol{\mathcal{X}}^{(i)}} \end{pmatrix}$$

$$= \begin{pmatrix} A \\ B \end{pmatrix} - \qquad [4.27]$$

$$\begin{pmatrix} a^{-1}_{s,\boldsymbol{\mathcal{X}}^{(i)}} q_{\boldsymbol{\mathcal{X}}^{(i)}} * \left(\sum_{j=0}^{2n} W^{(j)}_{(m)} h_A(\boldsymbol{\mathcal{X}}^{(j)}, \mathbf{u}_k) \right) * q^{-1}_{\boldsymbol{\mathcal{X}}^{(i)}} \\ b^{-1}_{s,\boldsymbol{\mathcal{X}}^{(i)}} q_{\boldsymbol{\mathcal{X}}^{(i)}} * \left(\sum_{j=0}^{2n} W^{(j)}_{(m)} h_B(\boldsymbol{\mathcal{X}}^{(j)}, \mathbf{u}_k) \right) * q^{-1}_{\boldsymbol{\mathcal{X}}^{(i)}} \end{pmatrix}$$

$$= \begin{pmatrix} A \\ B \end{pmatrix} - \sum_{j=0}^{2n} W^{(j)}_{(m)} \cdot$$

$$\begin{pmatrix} a^{-1}_{s,\boldsymbol{\mathcal{X}}^{(i)}} q_{\boldsymbol{\mathcal{X}}^{(i)}} * \left(a_{s,\boldsymbol{\mathcal{X}}^{(j)}} q^{-1}_{\boldsymbol{\mathcal{X}}^{(j)}} * A * q_{\boldsymbol{\mathcal{X}}^{(j)}} \right) * q^{-1}_{\boldsymbol{\mathcal{X}}^{(i)}} \\ b^{-1}_{s,\boldsymbol{\mathcal{X}}^{(i)}} q_{\boldsymbol{\mathcal{X}}^{(i)}} * \left(b_{s,\boldsymbol{\mathcal{X}}^{(j)}} q^{-1}_{\boldsymbol{\mathcal{X}}^{(j)}} * B * q_{\boldsymbol{\mathcal{X}}^{(j)}} \right) * q^{-1}_{\boldsymbol{\mathcal{X}}^{(i)}} \end{pmatrix}$$

$$= \begin{pmatrix} A \\ B \end{pmatrix} - \sum_{j=0}^{2n} W^{(j)}_{(m)} \cdot$$

$$\begin{pmatrix} a^{-1}_{s,\boldsymbol{\mathcal{X}}^{(i)}} a_{s,\boldsymbol{\mathcal{X}}^{(j)}} \cdot (q_{\boldsymbol{\mathcal{X}}^{(j)}} * q^{-1}_{\boldsymbol{\mathcal{X}}^{(i)}})^{-1} * A * (q_{\boldsymbol{\mathcal{X}}^{(j)}} * q^{-1}_{\boldsymbol{\mathcal{X}}^{(i)}}) \\ b^{-1}_{s,\boldsymbol{\mathcal{X}}^{(i)}} b_{s,\boldsymbol{\mathcal{X}}^{(j)}} \cdot (q_{\boldsymbol{\mathcal{X}}^{(j)}} * q^{-1}_{\boldsymbol{\mathcal{X}}^{(i)}})^{-1} * B * (q_{\boldsymbol{\mathcal{X}}^{(j)}} * q^{-1}_{\boldsymbol{\mathcal{X}}^{(i)}}) \end{pmatrix}$$

$$= \begin{pmatrix} A \\ B \end{pmatrix} - \sum_{j=0}^{2n} W^{(j)}_{(m)} \rho_{e-\eta(\boldsymbol{\mathcal{X}}^{(i)}, \boldsymbol{\mathcal{X}}^{(j)})} \begin{pmatrix} A \\ B \end{pmatrix}$$

$$= \sum_{j=0}^{2n} W^{(j)}_{(m)} \left[\begin{pmatrix} A \\ B \end{pmatrix} - \rho_{e-\eta(\boldsymbol{\mathcal{X}}^{(i)}, \boldsymbol{\mathcal{X}}^{(j)})} \begin{pmatrix} A \\ B \end{pmatrix} \right],$$

$$\text{since } \sum_{j=0}^{2n} W^{(j)}_{(m)} = 1. \qquad [4.28]$$

In equation [4.27], the notation $h_A(\cdot)$ and $h_B(\cdot)$ denotes the restriction of the observation model to the outputs associated with the acceleration and the magnetic field respectively. Equation [4.28] shows that the invariant prediction output errors can be written as a weighted sum of the (invariant) distances between the constant vector $(A^T \ B^T)^T$ and its image under the mapping ρ_g on the Lie group parametrized by the elements $e - \eta(\boldsymbol{\mathcal{X}}^{(i)}, \boldsymbol{\mathcal{X}}^{(j)})$, where i ranges over $[\![0\,; 2n]\!]$ and $\forall j \in [\![0\,; 2n]\!]$. This expression requires us to compute the values of the $(2n+1)$ invariant error vectors between the sigmas points $\eta(\boldsymbol{\mathcal{X}}^{(i)}, \boldsymbol{\mathcal{X}}^{(j)})$. As a special case, whenever $i \in [\![0\,; 2n]\!]$, then $\eta(\boldsymbol{\mathcal{X}}^{(i)}, \boldsymbol{\mathcal{X}}^{(i)}) = \vec{0}$. The algorithm therefore requires us to find $(2n+1) \times (2n+1) - (2n+1) = 4n^2 + 2n$ terms $\eta(\boldsymbol{\mathcal{X}}^{(i)}, \boldsymbol{\mathcal{X}}^{(j)})$ after these trivial cases are eliminated. Thus, in the simple case of an AHRS with $n = 9$, the SR-IUKF approach requires us to compute 342 invariant error vectors between the sigma points at each iteration. Equation [4.28] may be rewritten to eliminate the case $j = i$ in the weighted sum, giving, $\forall i \in [\![0\,; 2n]\!]$:

$$\begin{aligned}
\mathsf{E}(\hat{\mathbf{y}}, \boldsymbol{\mathcal{X}}^{(i)}, \hat{\mathbf{y}}^{(i)}) &= \sum_{j=0}^{2n} W_{(m)}^{(j)} \left[\begin{pmatrix} A \\ B \end{pmatrix} - \rho_{e-\eta(\boldsymbol{\mathcal{X}}^{(i)}, \boldsymbol{\mathcal{X}}^{(j)})} \begin{pmatrix} A \\ B \end{pmatrix} \right] \\
&= \sum_{\substack{j=0 \\ j \neq i}}^{2n} W_{(m)}^{(j)} \left[\begin{pmatrix} A \\ B \end{pmatrix} - \rho_{e-\eta(\boldsymbol{\mathcal{X}}^{(i)}, \boldsymbol{\mathcal{X}}^{(j)})} \begin{pmatrix} A \\ B \end{pmatrix} \right],
\end{aligned}$$

since

$$\begin{aligned}
\begin{pmatrix} A \\ B \end{pmatrix} - \rho_{e-\eta(\boldsymbol{\mathcal{X}}^{(i)}, \boldsymbol{\mathcal{X}}^{(i)})} \begin{pmatrix} A \\ B \end{pmatrix} &= \begin{pmatrix} A \\ B \end{pmatrix} - \rho_{e-\vec{0}} \begin{pmatrix} A \\ B \end{pmatrix} \\
&= \begin{pmatrix} A \\ B \end{pmatrix} - \rho_{e} \begin{pmatrix} A \\ B \end{pmatrix} = \vec{0}.
\end{aligned}$$

Combining the above calculations shows that, for the AHRS, given a set of $(2n+1)$ sigma points, we implicitly need to compute a potentially large number of invariant estimation errors between the sigma points in order to compute the invariant output errors. Figure 4.4 illustrates this for the AHRS and compares it with the more general approach presented at the beginning of this subsection (the generic (SR)-IUKF algorithm), which can be applied to any type of model.

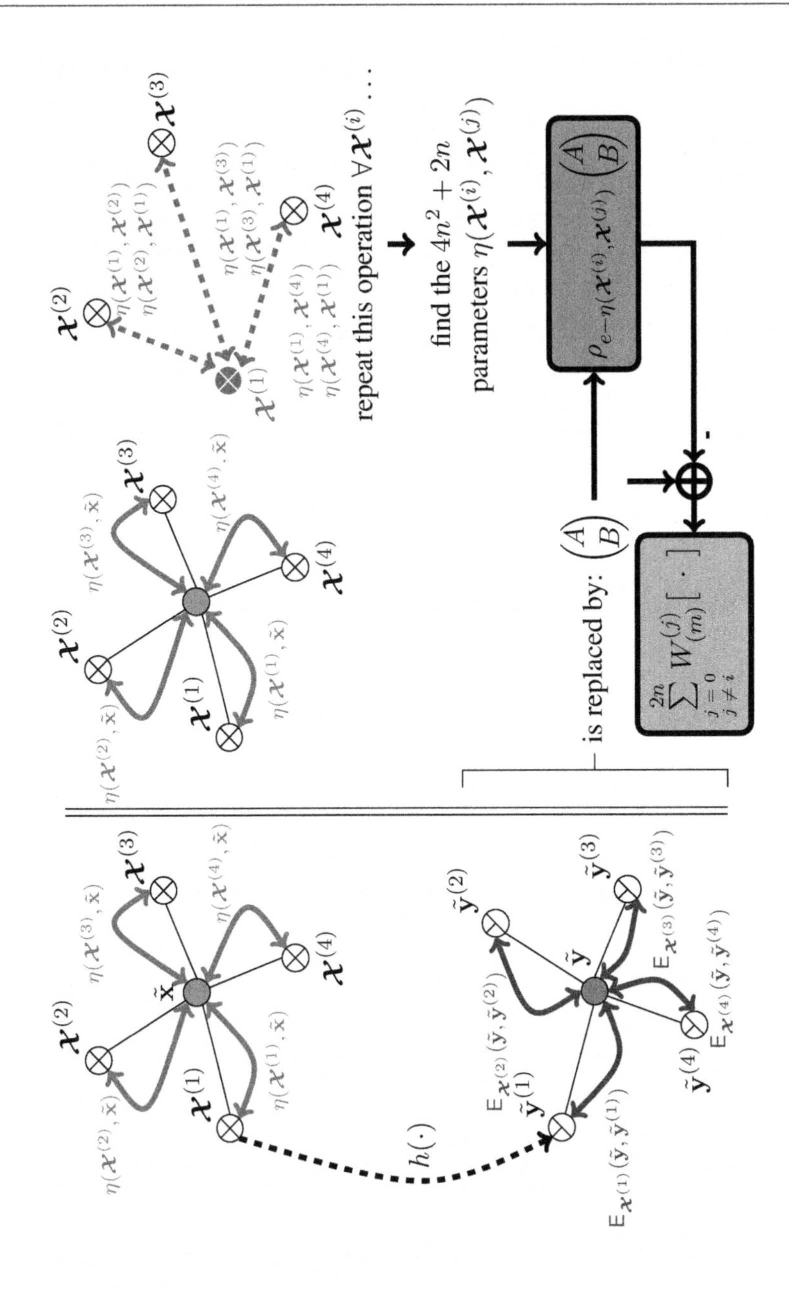

Figure 4.4. *Comparison of standard SR-IUKF and an applied version of SR-IUKF specifically adapted to the AHRS, parametrized by the sigma points* $\{\eta(\hat{\mathbf{x}}_{k+1|k}, \mathcal{X}^{(i)})/i \in [\![0\,; 2n]\!]\}$. *For a color version of this figure, see www.iste.co.uk/condomines/kalman.zip*

It is also worth noting that the parameter g of the composite group transformation $\phi_g = (\varphi_g, \psi_g, \rho_g)$, which here is a multiple parametrization, since each sigma point $\boldsymbol{\mathcal{X}}^{(i)}_{k+1|k}$, $i \in [\![0 ; 2n]\!]$ acts as a local parametrization when calculating the invariant errors, can also be chosen to be constant and equal to $\hat{\mathbf{x}}_{k+1|k}$, $\forall i \in [\![0 ; 2n]\!]$, without noticeably changing the performance of the estimation algorithm. This was confirmed by our first coded implementations of the (SR)-IUKF algorithm, which gave similar results in both cases. In the case where $g = \hat{\mathbf{x}}_{k+1|k}$, the invariant output prediction errors may be written as ($i \in [\![0 ; 2n]\!]$):

$$
\begin{aligned}
\mathsf{E}(\cdot) &= \rho_{\hat{\mathbf{x}}^{-1}_{k+1|k}}(\hat{\mathbf{y}}^{(i)}) - \rho_{\hat{\mathbf{x}}^{-1}_{k+1|k}}(\hat{\mathbf{y}}_{k+1|k}) \\
&= \rho_{\hat{\mathbf{x}}^{-1}_{k+1|k}}(h(\boldsymbol{\mathcal{X}}^{(i)}, \mathbf{u}_k)) - \rho_{\hat{\mathbf{x}}^{-1}_{k+1|k}}(\hat{\mathbf{y}}_{k+1|k}) \\
&= \rho_{\hat{\mathbf{x}}^{-1}_{k+1|k}}(h(\boldsymbol{\mathcal{X}}^{(i)}, \mathbf{u}_k)) - \sum_{j=0}^{2n} W^{(j)}_{(m)} \rho_{e-\eta(\hat{\mathbf{x}}_{k+1|k}, \boldsymbol{\mathcal{X}}^{(j)})} \begin{pmatrix} A \\ B \end{pmatrix} \\
&= h(\varphi_{\mathbf{x}^{-1}_{k+1|k}}(\boldsymbol{\mathcal{X}}^{(i)}), I(\hat{\mathbf{x}}_{k+1|k}, \mathbf{u}_k)) - \sum_{j=0}^{2n} W^{(j)}_{(m)} \rho_{e-\eta(\hat{\mathbf{x}}_{k+1|k}, \boldsymbol{\mathcal{X}}^{(j)})} \begin{pmatrix} A \\ B \end{pmatrix} \\
&= h(e - \eta(\hat{\mathbf{x}}_{k+1|k}, \boldsymbol{\mathcal{X}}^{(i)}), I(\hat{\mathbf{x}}_{k+1|k}, \mathbf{u}_k)) \\
&\quad - \sum_{j=0}^{2n} W^{(j)}_{(m)} \rho_{e-\eta(\hat{\mathbf{x}}_{k+1|k}, \boldsymbol{\mathcal{X}}^{(j)})} \begin{pmatrix} A \\ B \end{pmatrix}.
\end{aligned}
$$

From the equations $h(e - \eta(\hat{\mathbf{x}}_{k+1|k}, \boldsymbol{\mathcal{X}}^{(i)}), \psi_{\hat{\mathbf{x}}^{-1}_{k+1|k}}(\mathbf{u}_k))$ of the observation model of the AHRS, we also have that:

$$
h(\cdot) = \underbrace{\begin{pmatrix} a_{s,\boldsymbol{\mathcal{X}}^{(i)}} \cdot \hat{a}^{-1}_{s,k+1|k} \hat{q}_{k+1|k} * q^{-1}_{\boldsymbol{\mathcal{X}}^{(i)}} * A * q_{\boldsymbol{\mathcal{X}}^{(i)}} * \hat{q}^{-1}_{k+1|k} \\ b_{s,\boldsymbol{\mathcal{X}}^{(i)}} \hat{b}^{-1}_{s,k+1|k} \cdot \hat{q}_{k+1|k} * q^{-1}_{\boldsymbol{\mathcal{X}}^{(i)}} * B * q_{\boldsymbol{\mathcal{X}}^{(i)}} * \hat{q}^{-1}_{k+1|k} \end{pmatrix}}_{= \rho_{e-\eta(\hat{\mathbf{x}}_{k+1|k}, \boldsymbol{\mathcal{X}}^{(i)})} \begin{pmatrix} A \\ B \end{pmatrix}} . \qquad [4.29]
$$

Substituting the result of equation [4.29] into the previous formula gives:

$$\begin{aligned}\mathsf{E}(\hat{\mathbf{y}}, \hat{\mathbf{x}}_{k+1|k}, \hat{\mathbf{y}}^{(i)}) &= \rho_{e-\eta(\hat{\mathbf{x}}_{k+1|k}, \boldsymbol{\mathcal{X}}^{(i)})} \begin{pmatrix} A \\ B \end{pmatrix} \\ &\quad - \sum_{j=0}^{2n} W_{(m)}^{(j)} \rho_{e-\eta(\hat{\mathbf{x}}_{k+1|k}, \boldsymbol{\mathcal{X}}^{(j)})} \begin{pmatrix} A \\ B \end{pmatrix} \\ &= \sum_{\substack{j=0 \\ j \neq i}}^{2n} W_{(m)}^{(j)} \left[\rho_{e-\eta(\hat{\mathbf{x}}_{k+1|k}, \boldsymbol{\mathcal{X}}^{(i)})} \begin{pmatrix} A \\ B \end{pmatrix} \right. \\ &\quad \left. - \rho_{e-\eta(\hat{\mathbf{x}}_{k+1|k}, \boldsymbol{\mathcal{X}}^{(j)})} \begin{pmatrix} A \\ B \end{pmatrix} \right]. \end{aligned} \quad [4.30]$$

As before with the multiple parametrization, the calculation of the invariant output prediction errors reduces to a weighted sum of (invariant) output errors. The significance of this result is that each elementary error term takes an invariant of the estimation problem as an argument – namely the constant vector $(A^T \ B^T)^T$ – and that the parametrization of the Lie group ranges over the index j of the weighted sum, depending on the sigma point considered in each elementary calculation. Note that replacing $\hat{\mathbf{x}}_{k+1|k}$ by $\boldsymbol{\mathcal{X}}^{(i)}$ in equation [4.30] recovers exactly the same result as earlier. The advantage of this new parametrization is that equation [4.30] performs better computationally in the context of the (SR)-IUKF algorithm. Unlike the earlier case, we only need to know the invariant state estimation errors between the predicted state and each sigma point in order to determine the errors in the predicted outputs. As it happens, we already computed the set $\{\eta(\hat{\mathbf{x}}_{k+1|k}, \boldsymbol{\mathcal{X}}^{(i)})/i \in [\![0\,;2n]\!]\}$ during the prediction step to find the covariance matrix of the invariant state estimation errors. This formulation therefore avoids the need to find the $4n^2 + 2n$ invariant state errors between the sigma points. Figure 4.5 summarizes the computations of this alternatively parametrized formulation of the (SR)-IUKF and compares it with the generic version of the algorithm presented earlier.

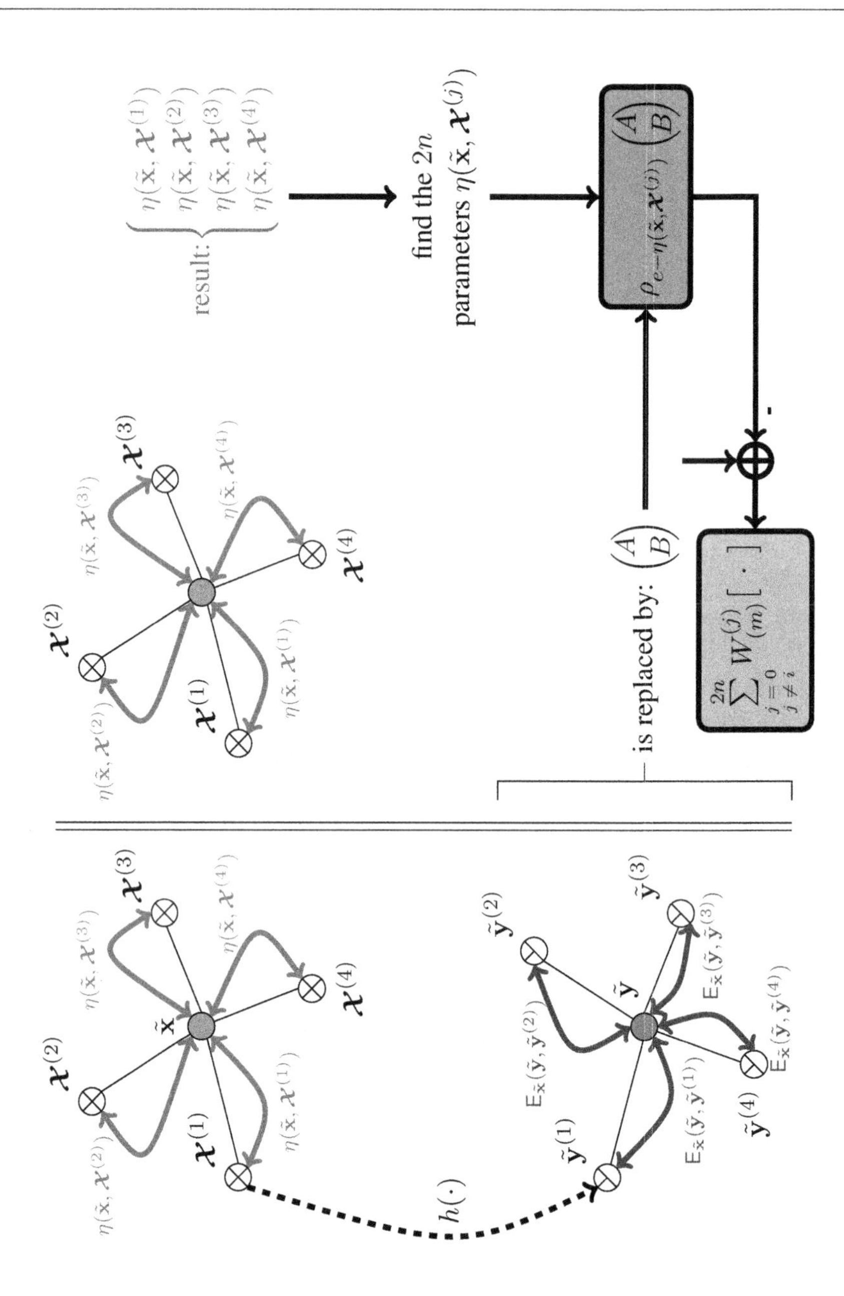

Figure 4.5. *Comparison of standard SR-IUKF and an applied variant of SR-IUKF specifically adapted to the AHRS, parametrized by the predicted state* $\hat{\mathbf{x}}_{k+1|k} = \sum_{j\in[\,0\,;2n\,]} W_{(m)}^{(j)}\, \boldsymbol{\mathcal{X}}_{k+1|k}^{(j)}$. *For a color version of this figure, see www.iste.co.uk/condomines/kalman.zip*

Equation [4.30] allows us to rewrite the IUKF equations of the AHRS problem as follows:

$$\hat{\mathbf{x}}_{k+1|k+1} = \hat{\mathbf{x}}_{k+1|k} + \sum_{i=1}^{n} \mathbf{K}_{k+1}^{(i)} \mathsf{E}(\hat{\mathbf{y}}_{k+1|k}, \hat{\mathbf{x}}_{k+1|k}, \mathbf{z}_{k+1}) \cdot w(\hat{\mathbf{x}}_{k+1|k})$$

$$= \hat{\mathbf{x}}_{k+1|k} + \sum_{i=1}^{n} \mathbf{K}_{k+1}^{(i)} (\rho_{\hat{\mathbf{x}}_{k+1|k}^{-1}}(\mathbf{z}_{k+1}) - \rho_{\hat{\mathbf{x}}_{k+1|k}^{-1}}(\hat{\mathbf{y}}_{k+1|k})) \cdot w_i(\hat{\mathbf{x}}_{k+1|k})$$

$$= \hat{\mathbf{x}}_{k+1|k} + \sum_{i=1}^{n} \mathbf{K}_{k+1}^{(i)} \left(\rho_{\hat{\mathbf{x}}_{k+1|k}^{-1}}(\mathbf{z}_{k+1}) - \sum_{j=0}^{2n} W_{(m)}^{(j)} \rho_{e-\eta(\hat{\mathbf{x}}_{k+1|k}, \boldsymbol{\mathcal{X}}^{(j)})} \begin{pmatrix} A \\ B \end{pmatrix} \right) \cdot w_i$$

$$= \hat{\mathbf{x}}_{k+1|k} + \sum_{i=1}^{n} \mathbf{K}_{k+1}^{(i)} \underbrace{\sum_{j=0}^{2n} W_{(m)}^{(j)} \left(\rho_{\hat{\mathbf{x}}_{k+1|k}^{-1}}(\mathbf{z}_{k+1}) - \rho_{e-\eta(\hat{\mathbf{x}}_{k+1|k}, \boldsymbol{\mathcal{X}}^{(j)})} \begin{pmatrix} A \\ B \end{pmatrix} \right)}_{\text{weighted sum of invariant innovation terms}} \cdot w_i$$

$$= \hat{\mathbf{x}}_{k+1|k} + \sum_{j=0}^{2n} W_{(m)}^{(j)} \sum_{i=1}^{n} \mathbf{K}_{k+1}^{(i)} \left(\rho_{\hat{\mathbf{x}}_{k+1|k}^{-1}}(\mathbf{z}_{k+1}) - \rho_{e-\eta(\hat{\mathbf{x}}_{k+1|k}, \boldsymbol{\mathcal{X}}^{(j)})} \begin{pmatrix} A \\ B \end{pmatrix} \right) \cdot w_i$$

$$= \sum_{j=0}^{2n} W_{(m)}^{(j)} \left[\boldsymbol{\mathcal{X}}_{k+1|k}^{(j)} + \sum_{i=1}^{n} \mathbf{K}_{k+1}^{(i)} \left(\rho_{\hat{\mathbf{x}}_{k+1|k}^{-1}}(\mathbf{z}_{k+1}) - \rho_{e-\eta(\hat{\mathbf{x}}_{k+1|k}, \boldsymbol{\mathcal{X}}^{(j)})} \begin{pmatrix} A \\ B \end{pmatrix} \right) \cdot w_i \right].$$

At each moment in time, the estimated state is therefore computed in the form of a correction of the prediction derived from the left or right invariant dynamics of the system, expressed as a weighted sum of invariant innovation terms on the Lie group parametrized directly or indirectly by $\hat{\mathbf{x}}_{k+1|k}$.

4.7.2. *Initial results and analysis*

The reference input data used for our first evaluation of an IUKF-type approach were generated by dynamic model simulations describing the free fall of a parafoil conducted as part of the FAWOPADS project, which studies control, guidance and navigation algorithms for automatic parachutes (see Figure 4.6). These simulated data provide a straightforward way to validate the methodological principles presented in this book, configure the parameters of each method and establish a few preliminary conclusions regarding the estimation algorithms presented above. We considered the data both with and without added noise. To make the various plots and figures shown below easier to read, the curves of the system state estimates are plotted without noise. A few results featuring added noise in the theoretical standard deviations and gain terms of the invariant Kalman filter are shown at the end of the chapter to demonstrate the effects of introducing noise (either pseudo-white Gaussian noise or colored noise) into the various measurements supplied to the estimator as inputs.

Figure 4.6. *MP-360 parafoil by Onyx with a payload*

The reference simulation that we used to validate our algorithms had a duration of slightly over 100 s. The simulated parachute system exhibits relatively strong dynamics, as can be seen in the time plots of the aircraft's reference attitude below. The roll, heading, and pitch angles vary by up to several dozen degrees. Although not shown here, the UAV also experienced significant variations in the velocity, partially invalidating one of the hypotheses of the model $\mathcal{M}_{ahrs}$, namely the assumption that $\dot{V} = 0$. It would therefore be interesting to investigate the effects of the error introduced into the estimation process by assuming that $\dot{V} = 0$. Analysis of the simulated data shows that $\dot{V}$ was non-zero throughout the period $t \in [5\,;40]$. The results and estimates obtained below by applying the (SR)-IUKF algorithm to the data are compared against results from the standard UKF algorithm in each case.

Estimate of the attitude

Figures 4.7 and 4.8 show the estimated attitude angles computed by the UKF and IUKF algorithms. Each estimate is compared against the pseudo-measurements reconstructed from the components of the reference quaternion state vector. In both cases, the estimated angles match the reference values almost perfectly. Both UKF and IUKF correctly reconstruct the attitude of the parafoil with respect to all three axes. Note that the error introduced into the initial state of the simulation was corrected very rapidly, after only a few computation steps (characteristic time < 0.5 s). However, the plots of the estimation errors (with a log scale along the vertical axis) show that the IUKF estimator converges more closely and quickly to the true values of the flight parameters. The IUKF achieves smaller estimation errors. Additionally, the comparison of these error plots suggests that the residuals of the state estimate constructed by IUKF appear to be more stable over time; a slight albeit slow decrease of these residuals may be observed in the results generated by the UKF algorithm. The estimates of the new IUKF approach therefore appear to be higher in quality than those delivered by the standard variant of the UKF algorithm considered here.

Estimate of the quaternion

Given that the estimates of the attitude angles were almost perfectly consistent with the reference values, it is not surprising that the same is true for each component of the quaternion. The estimated values computed by UKF and IUKF for the quaternion were close to the reference values at every point (see Figures 4.9 and 4.10). The ± 1 standard deviation lines (theoretical values, or in other words the values computed by the algorithms), plotted as dotted black lines for each component of the quaternion, are more interesting. In terms of the standard deviations, Figures 4.9 and 4.10 reveal clear differences between UKF and IUKF. Given identically configured estimated covariance matrices for the evolution and measurement noise terms ($\mathbf{Q}$ and $\mathbf{R}$), the uncertainty in the results produced by UKF fluctuates strongly, and possibly even diverges for the estimate of q_3. By contrast, after a transient phase of a few seconds, the theoretical standard deviations computed by IUKF converge to nearly constant values about the mean in each component $(\hat{q}_0, \hat{q}_1, \hat{q}_2, \hat{q}_3|\mathbf{z})$, assuming that the observations $\mathbf{z}$ are given. The effects of the inaccuracies in our choice of AHRS model (specifically, the fact that the dynamics of $\dot{V}$ are neglected) can be felt very slightly in the standard deviation plots, which fluctuate over the period $[5\,;40]$, albeit insignificantly.

These effects will be more visible in subsequent figures below, in particular Figure 4.13.

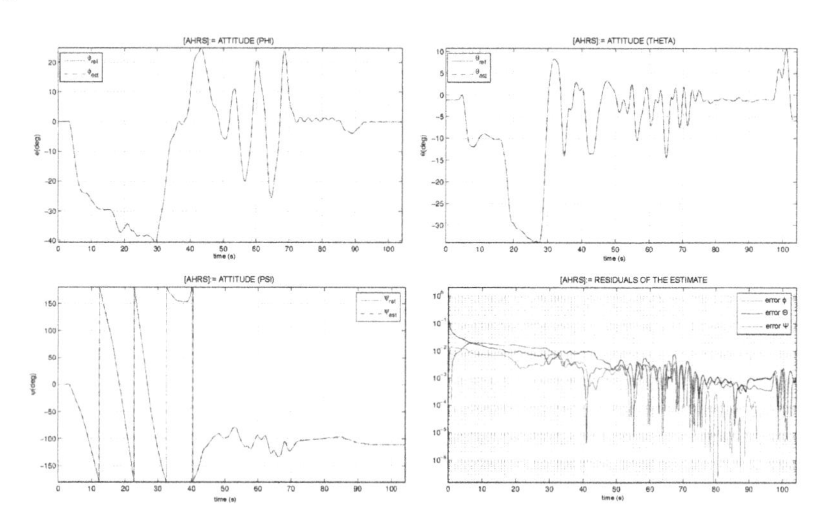

Figure 4.7. *UKF: estimates of the attitude angles (ϕ, θ, ψ) and estimation errors. For a color version of this figure, see www.iste.co.uk/condomines/kalman.zip*

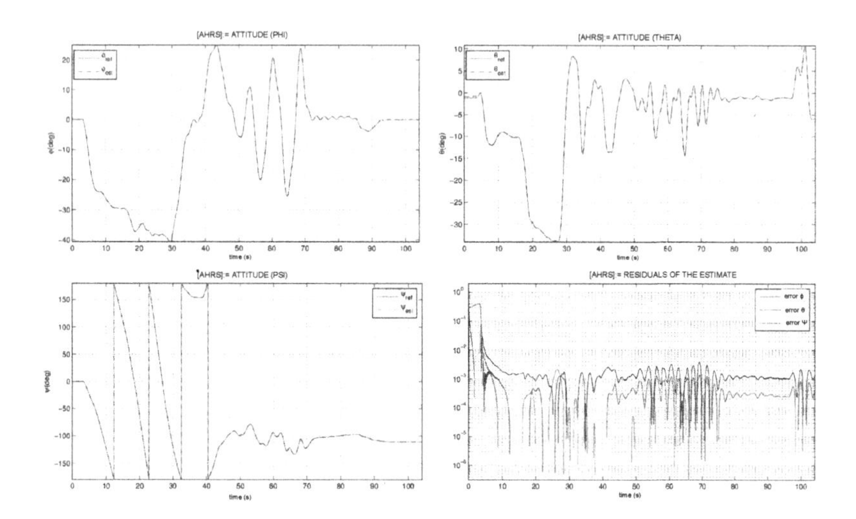

Figure 4.8. *IUKF: estimates of the attitude angles (ϕ, θ, ψ) and estimation errors. For a color version of this figure, see www.iste.co.uk/condomines/kalman.zip*

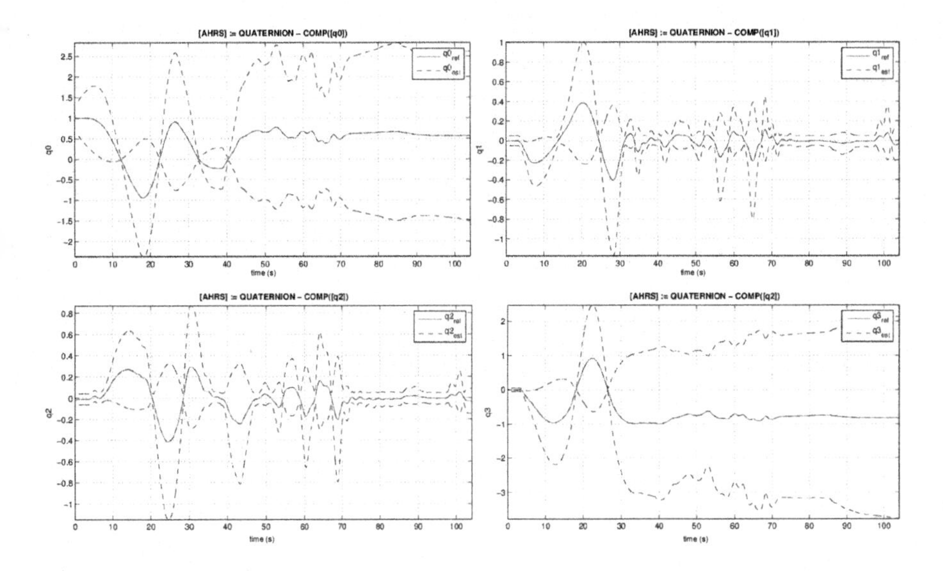

Figure 4.9. *UKF: estimates of the quaternion q and standard deviations. For a color version of this figure, see www.iste.co.uk/condomines/kalman.zip*

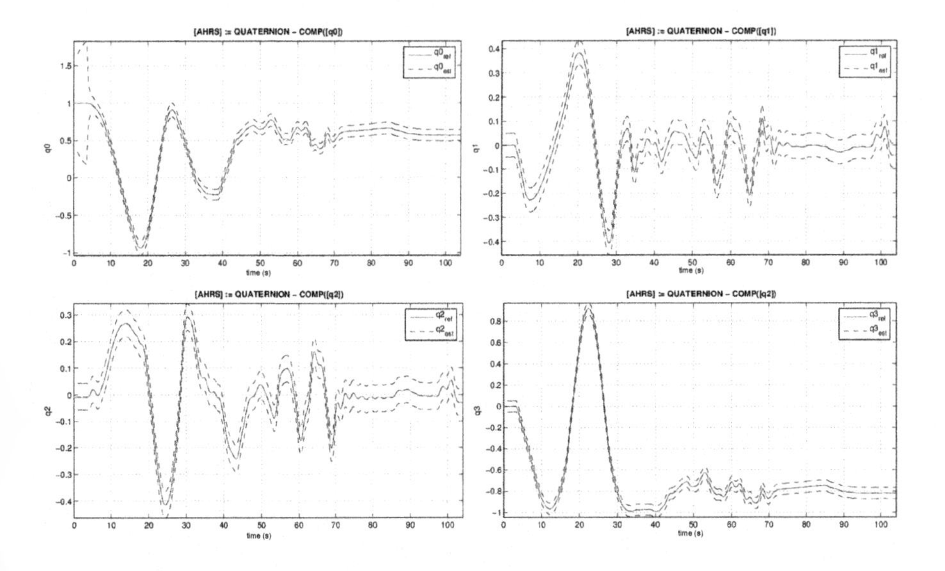

Figure 4.10. *IUKF: estimates of the quaternion q and standard deviations. For a color version of this figure, see www.iste.co.uk/condomines/kalman.zip*

Imperfections in the measurements

Three additive biases (represented by the state $\omega_b = (0.02\ 0.2\ \text{-}0.1)^T$) and two scale factors (represented by the states $(a_s, b_s) = (0.9, 1.2)$) were introduced in order to "pollute" the simulated data supplied to the estimation as inputs. This allows us to verify the observability of the estimation problem and test the ability of the algorithms to estimate these imperfections. The results are assembled in Figures 4.11 and 4.12. Both algorithms estimated the true values of the imperfections correctly, without error. The theoretical standard deviations calculated by UKF and its invariant counterpart are also almost identical. These results confirm that the problem is indeed observable, as concluded by the detectability/observability analysis performed in Chapter 3. As before in the components of the quaternion, a few parasitic effects are visible, caused by inaccuracies in the model used to process the data (which do not satisfy the assumption $\dot{V} = 0$). These effects can be seen in the theoretical standard deviations computed by IUKF for the states $\omega_b = (b\omega_p, b\omega_q, b\omega_r)$, which represent the biases in the gyrometers; they are more pronounced here than they were earlier, creating much more significant variations in the estimated standard deviations over time. The peaks of the dotted black curves coincide with the maximum values of $||\dot{V}||$. The estimates seem to converge slightly more quickly for IUKF than for UKF.

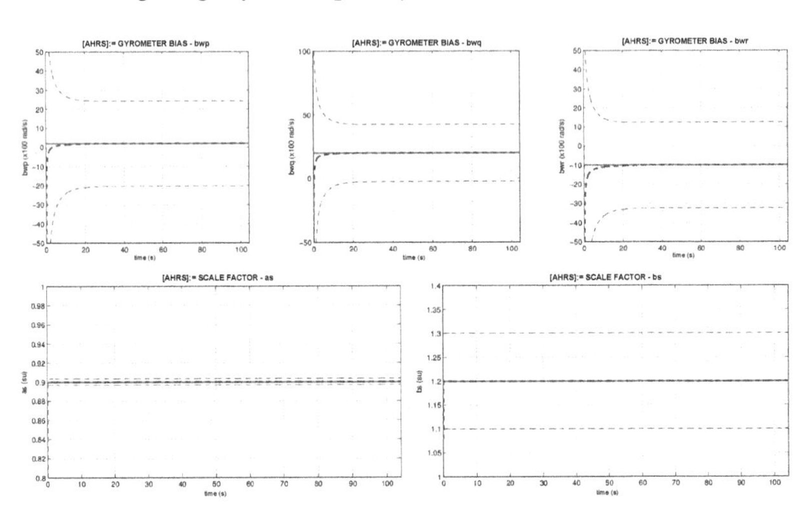

Figure 4.11. *UKF: bias ω_b and scale factors (a_s, b_s). For a color version of this figure, see www.iste.co.uk/condomines/kalman.zip*

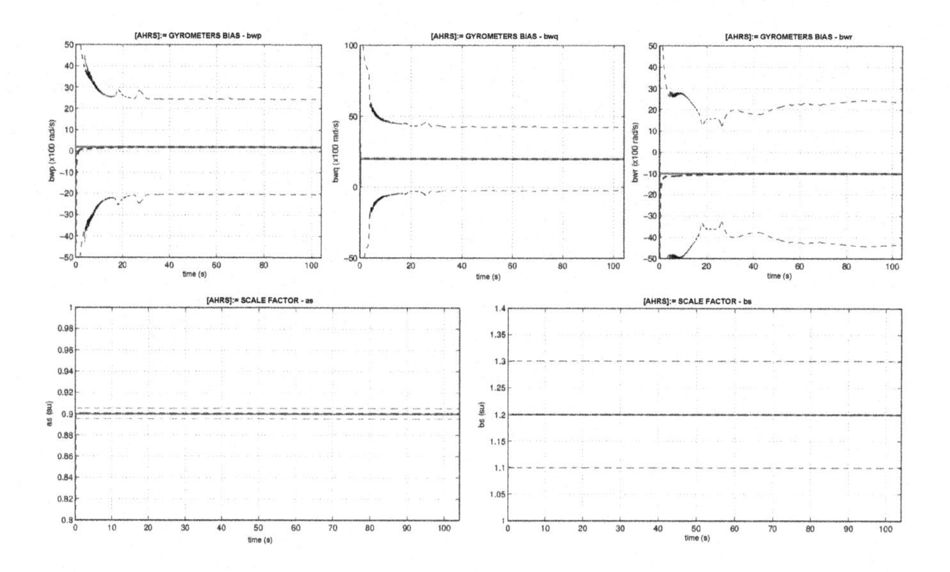

Figure 4.12. *IUKF: bias ω_b and scale factors (a_s, b_s). For a color version of this figure, see www.iste.co.uk/condomines/kalman.zip*

Comparison of the estimated covariances and the state estimation errors

Figure 4.13 compares the theoretical standard deviations computed by the UKF and IUKF algorithms for each component of the state vector. The results validate the theoretical principles discussed earlier, clearly showing that IUKF satisfies the properties of an invariant observer. The theoretical values of the standard deviation, and hence the estimated values of the covariance (as well as the gain terms), are constant on the trajectory defined by $I = \psi_{\hat{\mathbf{x}}^{-1}}(u = \omega_m)$, illustrating the invariant character of this estimation scheme. Thus, we have established an "equivalent" counterpart (IUKF) for the algorithm known as IEKF in the literature based on the principles of SUT, just as an "equivalent" version (UKF) was historically derived for the EKF algorithm. The results established for Kalman filters (regarding the connections between EKF and UKF) therefore continue to hold even in invariant settings. The invariance properties of the newly developed IUKF algorithm are also visible when we compare the state estimation errors. The values of these errors can be deduced from the simulation data. Figures 4.14 and 4.15 plot the norm of the estimation errors over time for each of the states q, ω_b, a_s and b_s. In these calculations, the initial state $\hat{\mathbf{x}}_0$ was varied for both algorithms, while keeping the norm of the initial state estimation error constant at all times. The results reveal the invariance properties of the IUKF algorithm; its behavior with regard to estimation errors is globally the same regardless of the estimate chosen for the initial condition, unlike the classical UKF algorithm. On every trajectory generated by varying the initial conditions, the dynamics of the convergence of the IUKF estimate to the true state were identical. Considering all of the results established above, it seems reasonable to conclude that the invariant Kalman filter proposed in this book is capable of accurately characterizing the uncertainty in the estimated state, in a similar spirit to the various other invariant observers that can already be found in the literature. The relatively constant nature of this uncertainty potentially makes it an extremely valuable source of information that could be exploited by various other problems (robust controls, FDIR, etc.).

Comparison of the correction gain terms

Figures 4.16 and 4.17 plot the values of the correction gain terms associated with each state over time for both the UKF and IUKF algorithms. For any given state, the vertical axes are scaled identically for both algorithms to make the results easier to read. Once again, the IUKF is characterized by the fact that the Kalman gain terms are remarkably close to constant, illustrating the effect of the invariant setting introduced to estimate the system state. The many frequent fluctuations in the UKF gain terms are almost completely eliminated by switching to the invariant version of the filter. Nonetheless, variations are visible in some of the IUKF correction gain terms (purple curve on the plots of q_0, $b\omega_p$, $b\omega_q$ and $b\omega_r$), likely caused by numerical errors and modeling inaccuracies (the hypothesis $\dot{V} = 0$ is not satisfied).

Results in the noisy case

Figures 4.18–4.20 compare the results produced by UKF and IUKF when the simulated data includes noise. Here, a series of additive colored noise terms were incorporated into the reference simulation as perturbations of the measurements ω_m, y_{Am}, and y_{Bm}. Figure 4.18 compares the theoretical standard deviations computed by each method. Figures 4.19 and 4.20 provide some perspective for the gain terms. The same conclusions can be drawn as in the noiseless case. These figures also show that IUKF is extremely good at filtering out high-frequency perturbations by comparison with UKF.

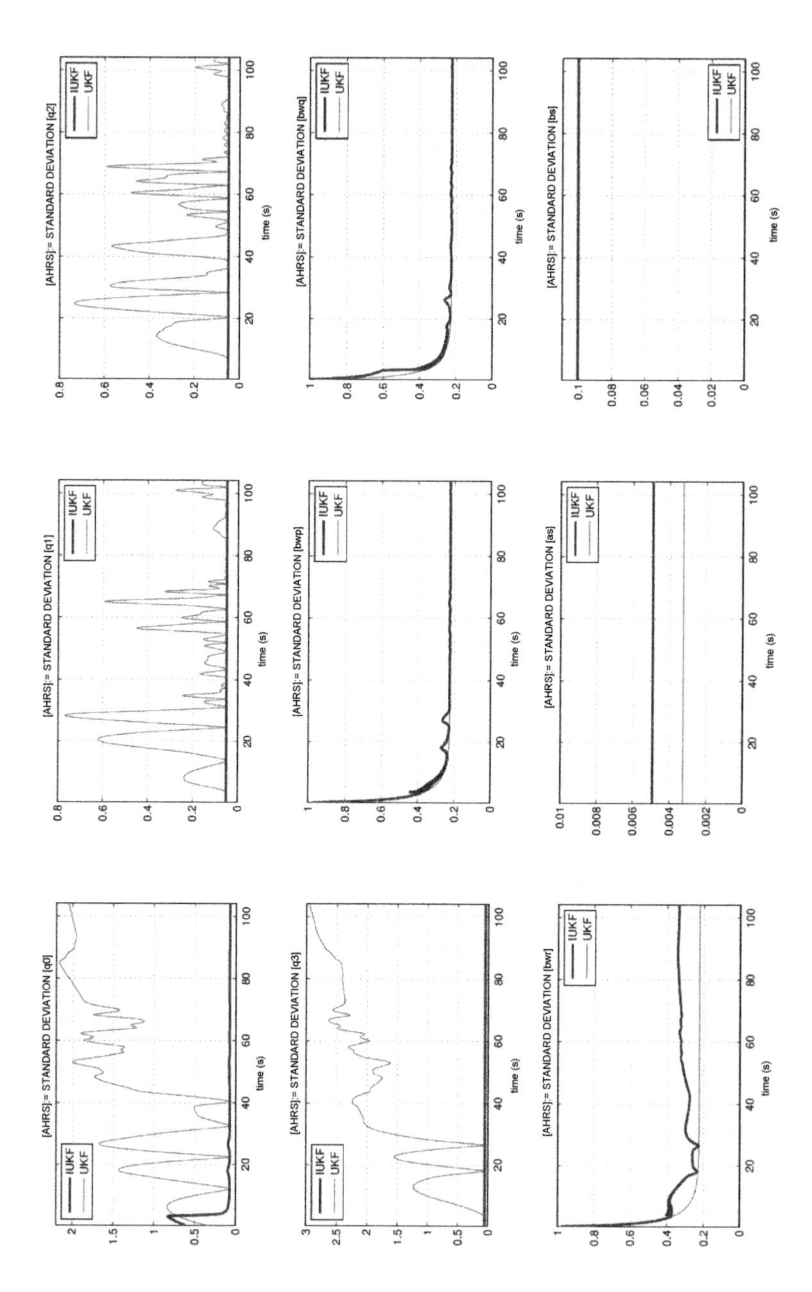

Figure 4.13. *Comparison of the theoretical standard deviations computed by the UKF and IUKF algorithms. For a color version of this figure, see www.iste.co.uk/condomines/kalman.zip*

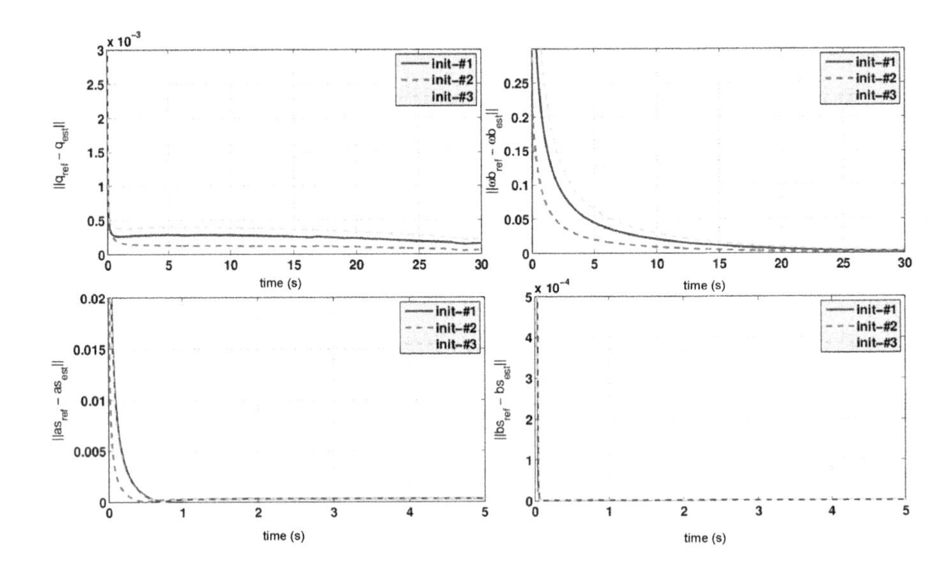

Figure 4.14. *UKF: state estimation error* $\mathbf{x} - \hat{\mathbf{x}}$. *For a color version of this figure, see www.iste.co.uk/condomines/kalman.zip*

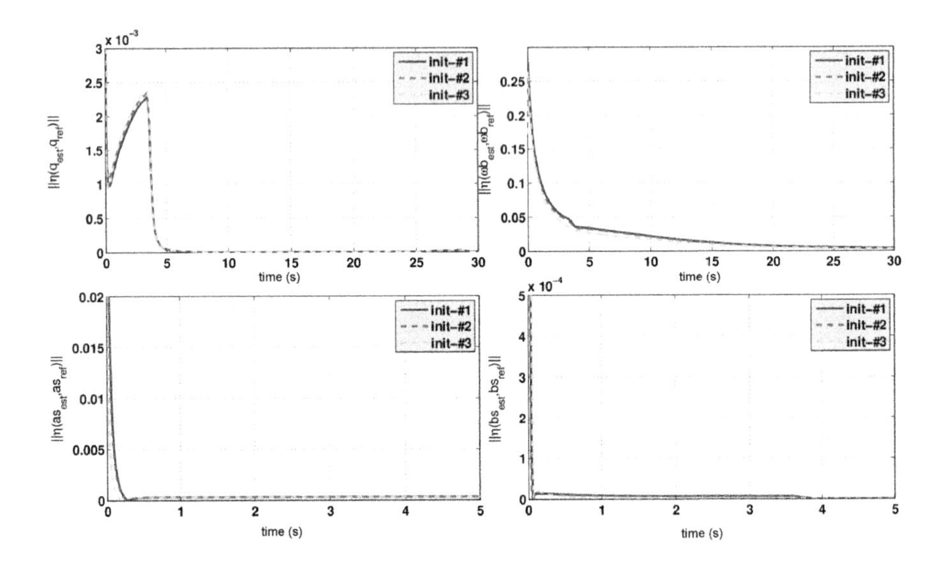

Figure 4.15. *IUKF: invariant state estimation error* $\eta(\mathbf{x}, \hat{\mathbf{x}})$. *For a color version of this figure, see www.iste.co.uk/condomines/kalman.zip*

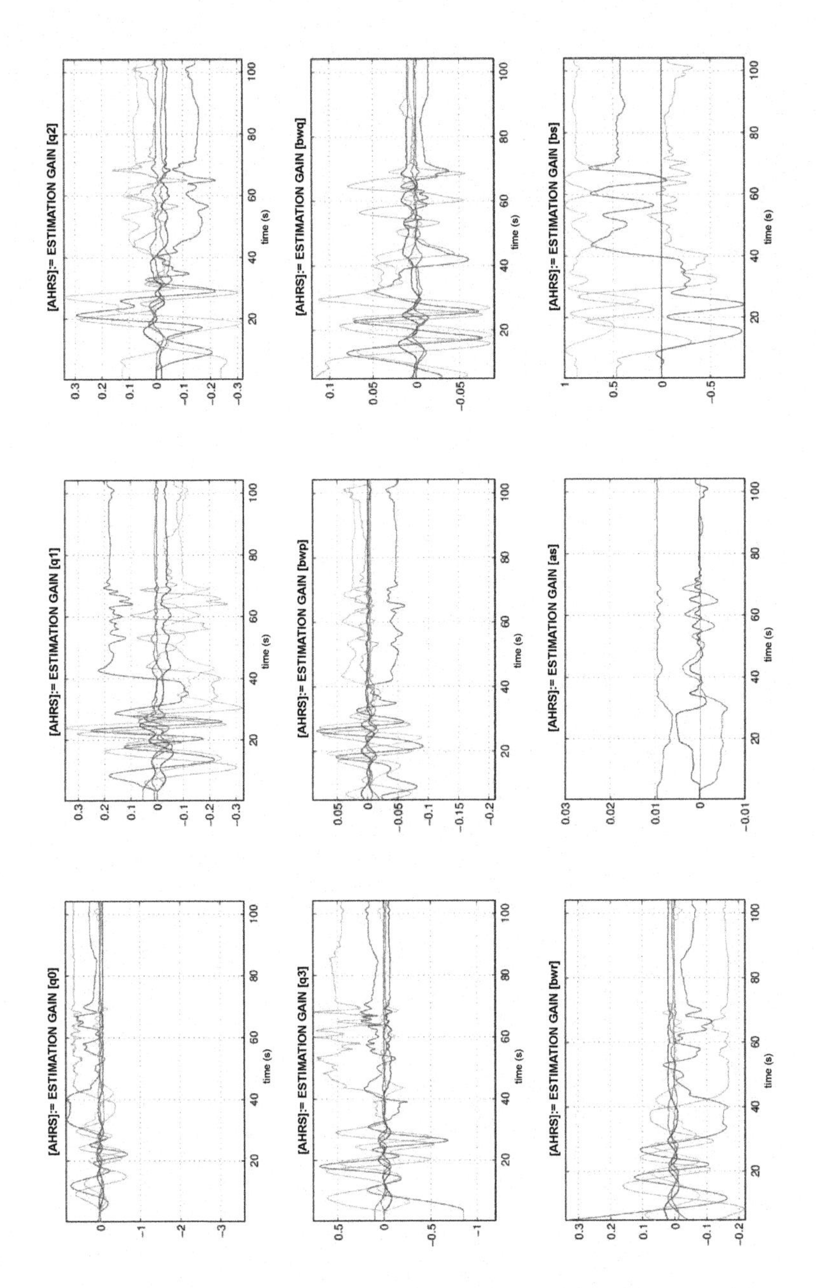

Figure 4.16. *UKF: calculated values of the correction gains. For a color version of this figure, see www.iste.co.uk/condomines/kalman.zip*

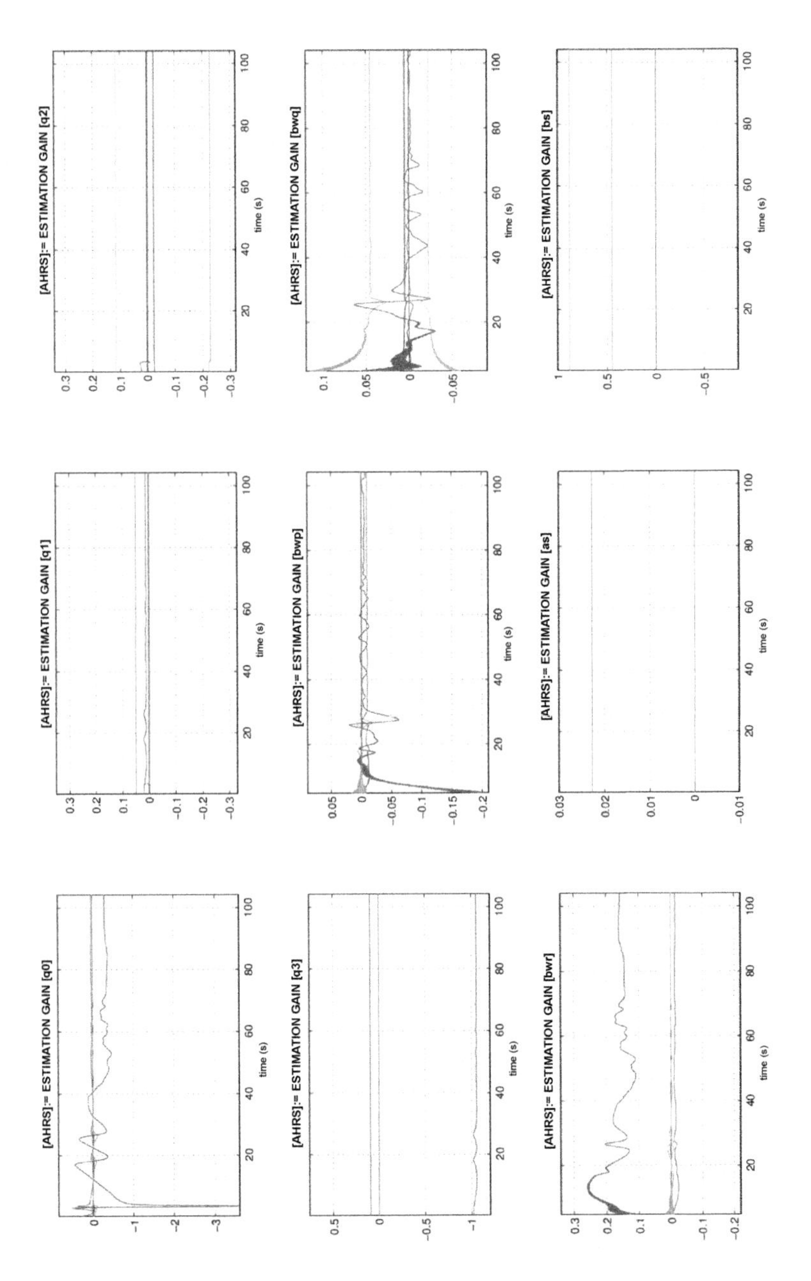

Figure 4.17. *IUKF: calculated value of the correction gains. For a color version of this figure, see www.iste.co.uk/condomines/kalman.zip*

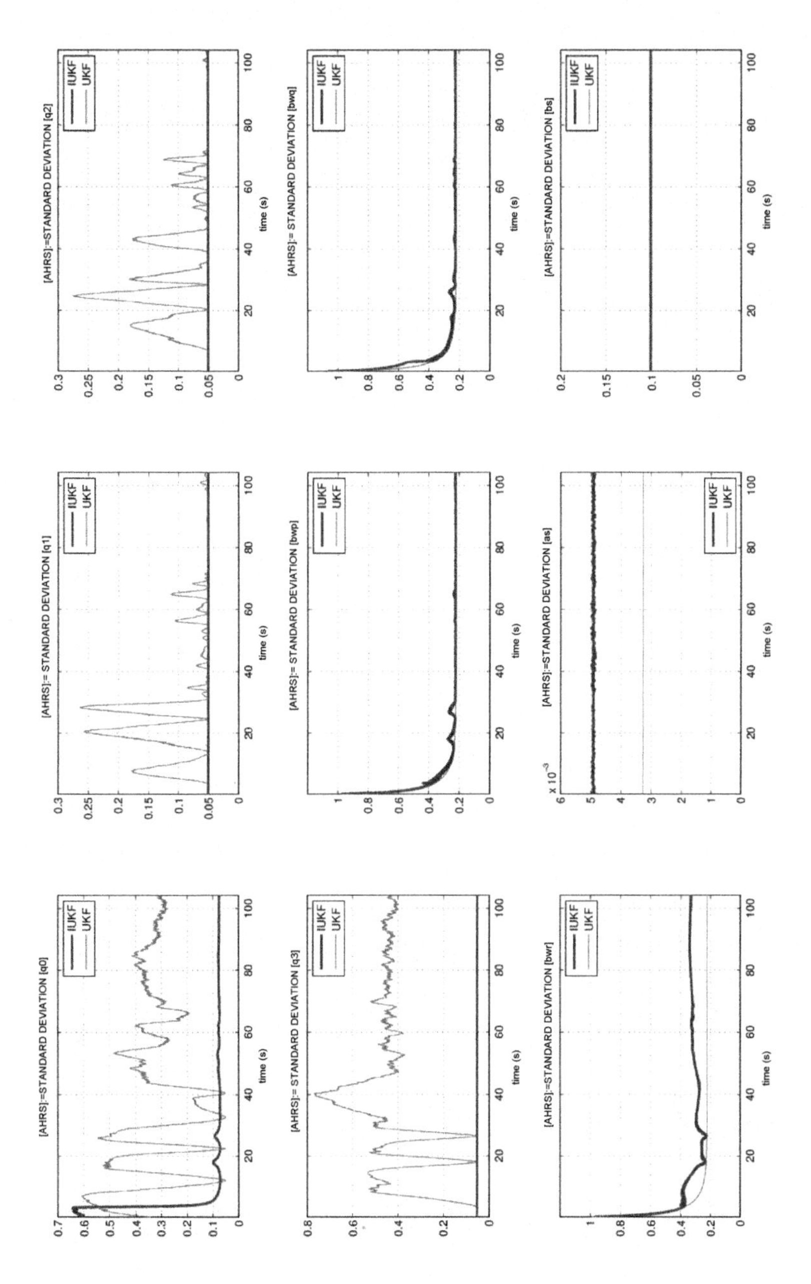

Figure 4.18. *Noisy case – comparison of the theoretical standard deviations of UKF and IUKF. For a color version of this figure, see www.iste.co.uk/condomines/kalman.zip*

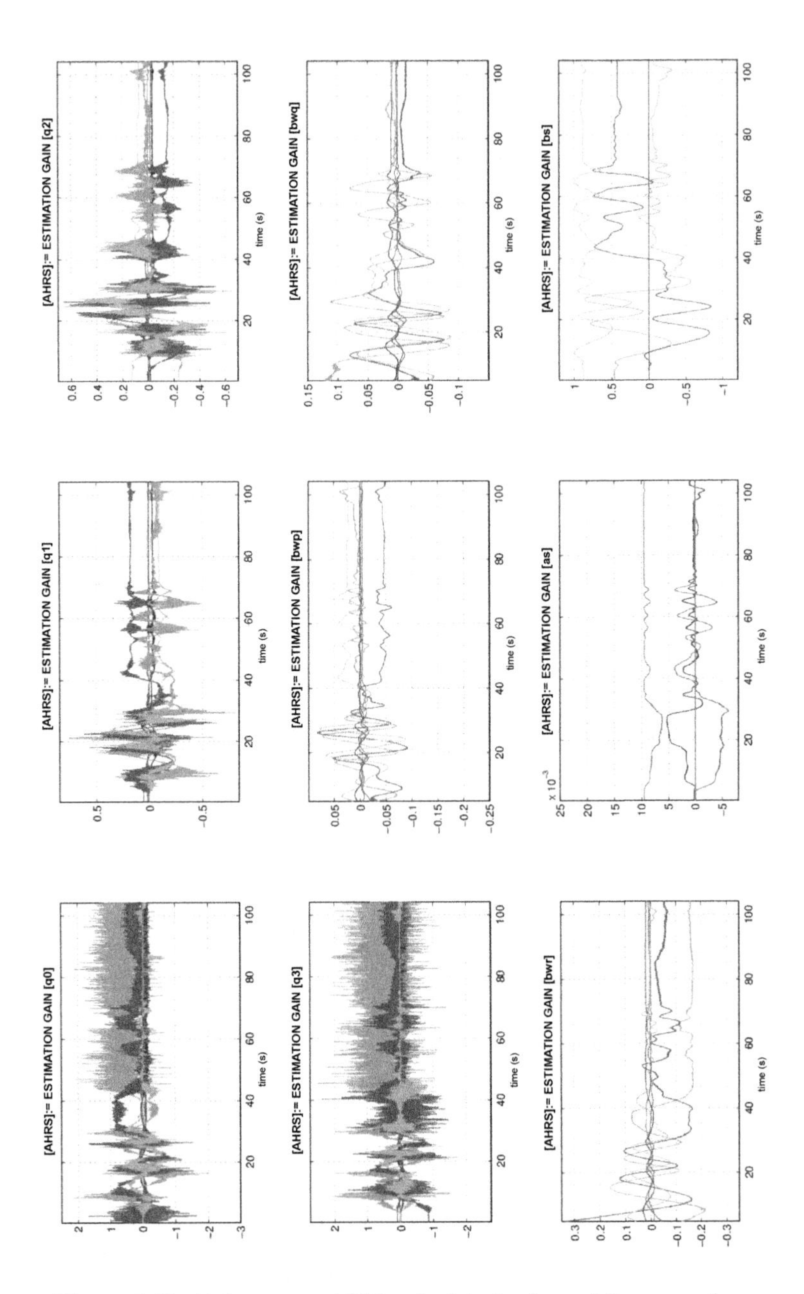

Figure 4.19. *Noisy case – UKF: calculated values of the correction gains. For a color version of this figure, see www.iste.co.uk/condomines/kalman.zip*

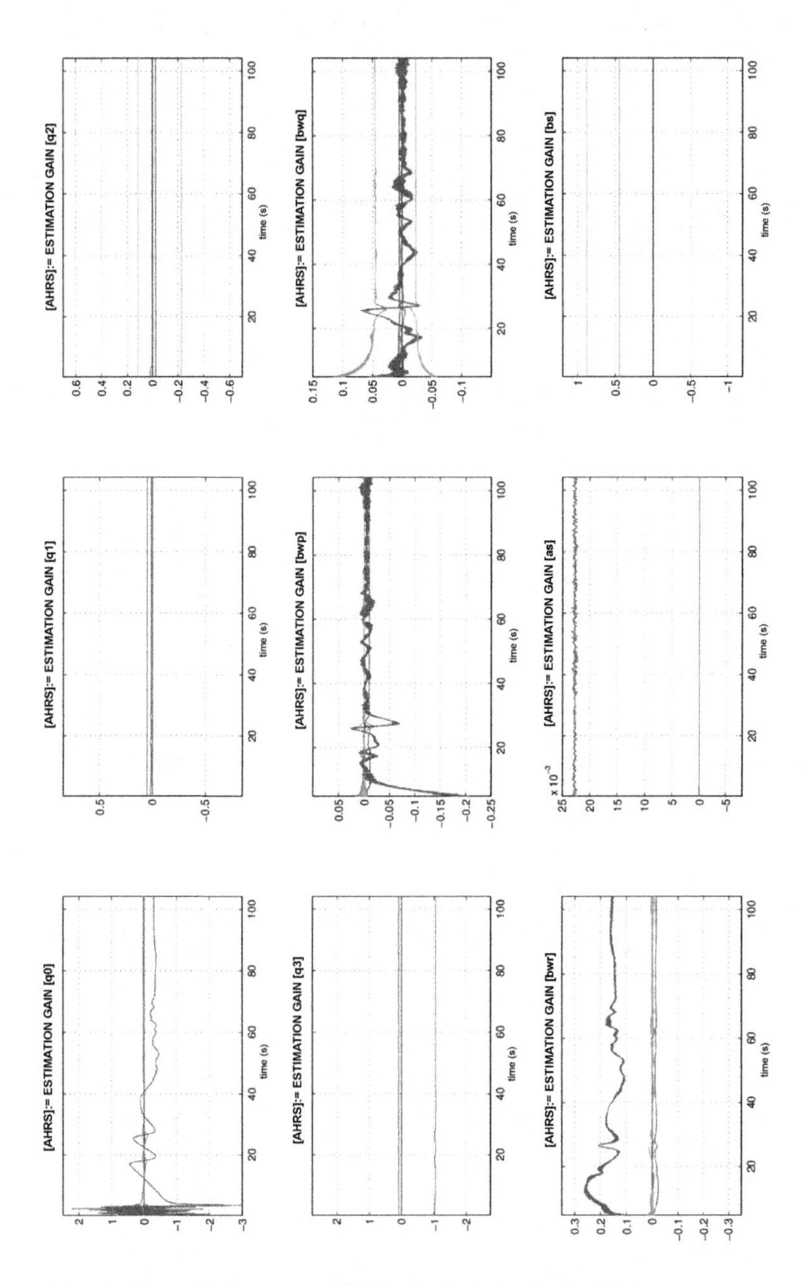

Figure 4.20. *Noisy case – IUKF: calculated values of the correction gains. For a color version of this figure, see www.iste.co.uk/condomines/kalman.zip*

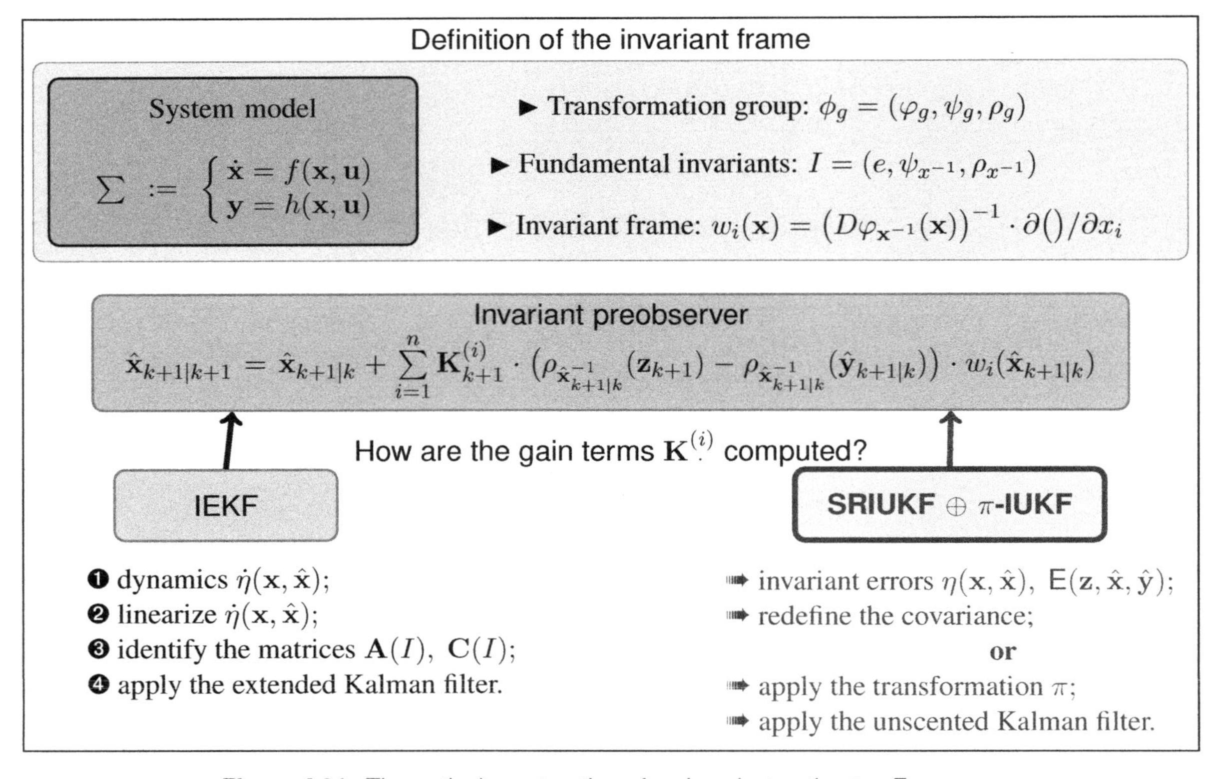

Figure 4.21. *Theoretical construction of an invariant estimator. For a color version of this figure, see www.iste.co.uk/condomines/kalman.zip*

4.7.3. *Summary and pseudocode*

Figure 4.21 summarizes the approach to constructing invariant state observers followed throughout these pages. The diagram distinguishes between:

– the *extended model* of the system dynamics, which includes the definition of the state representation, but also defines the transformation group. Based on these transformations, we can then deduce the fundamental invariants and the invariant frame of the system;

– the *definition of the estimating filter*, also known as the candidate invariant pre-observer, and the definition of its structure, phrased in terms of the fundamental invariants of the system and the notion of invariant frame;

– finally, the *computation of the correction gain terms* that appear in the structural equations of the estimator. These gain terms can be calculated using classical techniques inspired by Kalman filtering. Recent work in this field suggests that it is possible to compute these gain terms by adapting an extended Kalman filter to the invariant setting at the cost of linearizing the invariant state estimation error. In this book, we propose to tackle the problem of computing the gain terms by unscented Kalman filtering; two algorithmic solutions are proposed: (SR)-IUKF and π-(SR)IUKF.

The pseudocode for the IUKF algorithm and its factorized variant, SR-IUKF, is listed as follows.

IUKF algorithm for estimating the state with additive noise terms

Description of the procedure: $\mathbf{x}_{k+1} = f(\mathbf{x}_k, \mathbf{u}_k) + \mathbf{w}_k$

Model of the sensors: $\mathbf{y}_k = h(\mathbf{x}_k, \mathbf{u}_k) + \mathbf{v}_k$

Find the composite group transformation: $\phi_g = (\varphi_g, \psi_g, \rho_g)$

- Compute the weights $W_m^{(j)}$, $W_c^{(j)}$, $j \in [\![0 ; 2n]\!]$ (see Scaled UKF)
- Initialize $\hat{\mathbf{x}}_0 = \mathsf{E}[\mathbf{x}_0]$ and $\hat{\mathbf{P}}_0 = \mathsf{E}[\eta(\mathbf{x}_0, \hat{\mathbf{x}}_0)\eta^T(\mathbf{x}_0, \hat{\mathbf{x}}_0)]$

- Compute the $(2n+1)$ sigma points: $\boldsymbol{\mathcal{X}}_{k|k}^{(0)} = \hat{\mathbf{x}}_{k|k}$, $\boldsymbol{\mathcal{X}}_{k|k}^{(1)}$, $\ldots$, $\boldsymbol{\mathcal{X}}_{k|k}^{(2n)}$

- Prediction step:

❶ $\forall i \in [\![0 ; 2n]\!]$, $\boldsymbol{\mathcal{X}}_{k+1|k}^{(i)} = f(\boldsymbol{\mathcal{X}}_{k|k}^{(i)}, \mathbf{u}_k) \Rightarrow \hat{\mathbf{x}}_{k+1|k} = \sum_{i=0}^{2n} W_{(m)}^{(i)} \boldsymbol{\mathcal{X}}_{k+1|k}^{(i)}$

❷ Predict the covariance of the invariant state estimation errors:

$$\hat{\mathbf{P}}_{\mathbf{xx},k+1|k} = \begin{cases} \mathbf{Q} + \sum_{i=0}^{2n} W_{(c)}^{(i)} \eta(\boldsymbol{\mathcal{X}}_{k+1|k}^{(i)}, \hat{\mathbf{x}}_{k+1|k})\eta^T(\boldsymbol{\mathcal{X}}_{k+1|k}^{(i)}, \hat{\mathbf{x}}_{k+1|k}) & \rightsquigarrow \text{ parameters } \boldsymbol{\mathcal{X}}^{(i)} \\ \mathbf{Q} + \sum_{i=0}^{2n} W_{(c)}^{(i)} \eta(\hat{\mathbf{x}}_{k+1|k}, \boldsymbol{\mathcal{X}}_{k+1|k}^{(i)})\eta^T(\hat{\mathbf{x}}_{k+1|k}, \boldsymbol{\mathcal{X}}_{k+1|k}^{(i)}) & \rightsquigarrow \text{ parameter } \hat{\mathbf{x}}_{k+1|k} \end{cases}$$

❸ $\forall i \in [\![0 ; 2n]\!]$, $\hat{\mathbf{y}}_{k+1|k}^{(i)} = h(\boldsymbol{\mathcal{X}}_{k+1|k}^{(i)}, \mathbf{u}_k) \Rightarrow \hat{\mathbf{y}}_{k+1|k} = \sum_{i=0}^{2n} W_{(m)}^{(i)} \hat{\mathbf{y}}_{k+1|k}^{(i)}$

❹ Predict the covariance of the invariant output estimation errors:

$$\hat{\mathbf{P}}_{\mathbf{yy},k+1|k} = \begin{cases} \mathbf{R} + \sum_{i=0}^{2n} W_{(c)}^{(i)} \mathsf{E}(\hat{\mathbf{y}}_{k+1|k}, \boldsymbol{\mathcal{X}}_{k+1|k}^{(i)}, \hat{\mathbf{y}}_{k+1|k}^{(i)})\mathsf{E}^T(\hat{\mathbf{y}}_{k+1|k}, \boldsymbol{\mathcal{X}}_{k+1|k}^{(i)}, \hat{\mathbf{y}}_{k+1|k}^{(i)}) \\ \mathbf{R} + \sum_{i=0}^{2n} W_{(c)}^{(i)} \mathsf{E}(\hat{\mathbf{y}}_{k+1|k}, \hat{\mathbf{x}}_{k+1|k}, \hat{\mathbf{y}}_{k+1|k}^{(i)})\mathsf{E}^T(\hat{\mathbf{y}}_{k+1|k}, \hat{\mathbf{x}}_{k+1|k}, \hat{\mathbf{y}}_{k+1|k}^{(i)}) \end{cases}$$

❺ Predict the cross-covariance of the invariant estimation errors:

$$\hat{\mathbf{P}}_{\mathbf{xy},k+1|k} = \begin{cases} \sum_{i=0}^{2n} W_{(c)}^{(i)} \eta(\boldsymbol{\mathcal{X}}_{k+1|k}^{(i)}, \hat{\mathbf{x}}_{k+1|k})\mathsf{E}^T(\hat{\mathbf{y}}_{k+1|k}, \boldsymbol{\mathcal{X}}_{k+1|k}^{(i)}, \hat{\mathbf{y}}_{k+1|k}^{(i)}) \\ \sum_{i=0}^{2n} W_{(c)}^{(i)} \eta(\hat{\mathbf{x}}_{k+1|k}, \boldsymbol{\mathcal{X}}_{k+1|k}^{(i)})\mathsf{E}^T(\hat{\mathbf{y}}_{k+1|k}, \hat{\mathbf{x}}_{k+1|k}, \hat{\mathbf{y}}_{k+1|k}^{(i)}) \end{cases}$$

- Correction/filtering step:

❶ Compute the gain terms: $\mathbf{K}_{k+1} = \hat{\mathbf{P}}_{\mathbf{xy},k+1|k} \cdot \hat{\mathbf{P}}_{\mathbf{yy},k+1|k}^{-1}$

❷ Compute the new estimated state:

$$\hat{\mathbf{x}}_{k+1|k+1} = \hat{\mathbf{x}}_{k+1|k} + \sum_{i=1}^{n} \mathbf{K}_{k+1}^{(i)} \cdot \mathsf{E}(\hat{\mathbf{y}}_{k+1|k}, \hat{\mathbf{x}}_{k+1|k}, \mathbf{z}_{k+1}) \cdot w(\hat{\mathbf{x}}_{k+1|k})$$

❸ Update the covariance $\hat{\mathbf{P}}_{\mathbf{xx}}$:

$$\hat{\mathbf{P}}_{\mathbf{xx},k+1|k+1} = \hat{\mathbf{P}}_{\mathbf{xx},k+1|k} - \mathbf{K}_{k+1}\hat{\mathbf{P}}_{\mathbf{zz},k+1|k}\mathbf{K}_{k+1}^T$$

SR-IUKF algorithm for estimating the state with additive noise terms
Description of the procedure: $\mathbf{x}_{k+1} = f(\mathbf{x}_k, \mathbf{u}_k) + \mathbf{w}_k$ Model of the sensors: $\mathbf{y}_k = h(\mathbf{x}_k, \mathbf{u}_k) + \mathbf{v}_k$ Find the composite transformation group: $\phi_g = (\varphi_g, \psi_g, \rho_g)$
► Compute the weights $W_m^{(j)}$, $W_c^{(j)}$, $j \in [\![0 ; 2n]\!]$ (see Scaled UKF)
► Initialize $\hat{\mathbf{x}}_0 = \mathsf{E}[\mathbf{x}_0]$ and $\hat{\mathbf{P}}_0 = \mathsf{E}[\eta(\mathbf{x}_0, \hat{\mathbf{x}}_0)\eta^T(\mathbf{x}_0, \hat{\mathbf{x}}_0)]$

► Compute the $(2n+1)$ sigma points: $\boldsymbol{\mathcal{X}}_{k\|k}^{(0)} = \hat{\mathbf{x}}_{k\|k}, \boldsymbol{\mathcal{X}}_{k\|k}^{(1)}, \ldots, \boldsymbol{\mathcal{X}}_{k\|k}^{(2n)}$

► Prediction step:

❶ $\forall i \in [\![0 ; 2n]\!]$, $\boldsymbol{\mathcal{X}}_{k+1|k}^{(i)} = f(\boldsymbol{\mathcal{X}}_{k|k}^{(i)}, \mathbf{u}_k) \Rightarrow \hat{\mathbf{x}}_{k+1|k} = \sum_{i=0}^{2n} W_{(m)}^{(i)} \boldsymbol{\mathcal{X}}_{k+1|k}^{(i)}$

❷ Find the QR factorization of the predicted covariance of the invariant state errors:

$$\hat{\mathbf{S}}_{\mathbf{xx},k+1|k} = \begin{cases} ① \ \mathsf{qr}\left[\sqrt{W_{(c)}^{(1)}}\left(\eta(\boldsymbol{\mathcal{X}}_{k+1|k}^{(1)}, \hat{\mathbf{x}}_{k+1|k}) \ldots \eta(\boldsymbol{\mathcal{X}}_{k+1|k}^{(2n)}, \hat{\mathbf{x}}_{k+1|k})\right) \ \mathbf{Q}^{1/2}\right] \\ ② \ \mathsf{cholupdate}\left(\hat{\mathbf{S}}_{\mathbf{xx},k+1|k}, \eta(\boldsymbol{\mathcal{X}}_{k+1|k}^{(0)}, \hat{\mathbf{x}}_{k+1|k}), W_{(c)}^{(0)}\right) \end{cases} \rightsquigarrow \text{ parameters } \boldsymbol{\mathcal{X}}^{(i)}$$

$$\hat{\mathbf{S}}_{\mathbf{xx},k+1|k} = \begin{cases} ① \ \mathsf{qr}\left[\sqrt{W_{(c)}^{(1)}}\left(\eta(\hat{\mathbf{x}}_{k+1|k}, \boldsymbol{\mathcal{X}}_{k+1|k}^{(1)}) \ldots \eta(\hat{\mathbf{x}}_{k+1|k}, \boldsymbol{\mathcal{X}}_{k+1|k}^{(2n)})\right) \ \mathbf{Q}^{1/2}\right] \\ ② \ \mathsf{cholupdate}\left(\hat{\mathbf{S}}_{\mathbf{xx},k+1|k}, \eta(\hat{\mathbf{x}}_{k+1|k}, \boldsymbol{\mathcal{X}}_{k+1|k}^{(0)}), W_{(c)}^{(0)}\right) \end{cases} \rightsquigarrow \text{ parameter } \hat{\mathbf{x}}_{k+1|k}$$

❸ $\forall i \in [\![0 ; 2n]\!]$, $\hat{\mathbf{y}}_{k+1|k}^{(i)} = h(\boldsymbol{\mathcal{X}}_{k+1|k}^{(i)}, \mathbf{u}_k) \Rightarrow \hat{\mathbf{y}}_{k+1|k} = \sum_{i=0}^{2n} W_{(m)}^{(i)} \hat{\mathbf{y}}_{k+1|k}^{(i)}$

❹ Find the QR factorization of the predicted covariance of the invariant output errors:

$$\hat{\mathbf{S}}_{\mathbf{yy},k+1|k} = \begin{cases} ① \ \mathsf{qr}\left[\sqrt{W_{(c)}^{(1)}}\left(\mathsf{E}(\hat{\mathbf{y}}_{k+1|k}, \boldsymbol{\mathcal{X}}_{k+1|k}^{(1)}, \hat{\mathbf{y}}_{k+1|k}^{(1)}) \ldots \mathsf{E}(\hat{\mathbf{y}}_{k+1|k}, \boldsymbol{\mathcal{X}}_{k+1|k}^{(2n)}, \hat{\mathbf{y}}_{k+1|k}^{(2n)})\right) \ \mathbf{R}^{1/2}\right] \\ ② \ \mathsf{cholupdate}\left(\hat{\mathbf{S}}_{\mathbf{yy},k+1|k}, \mathsf{E}(\hat{\mathbf{y}}_{k+1|k}, \boldsymbol{\mathcal{X}}_{k+1|k}^{(0)}, \hat{\mathbf{y}}_{k+1|k}^{(0)}), W_{(c)}^{(0)}\right) \end{cases}$$

$$\hat{\mathbf{S}}_{\mathbf{yy},k+1|k} = \begin{cases} ① \ \mathsf{qr}\left[\sqrt{W_{(c)}^{(1)}}\left(\mathsf{E}(\hat{\mathbf{y}}_{k+1|k}, \hat{\mathbf{x}}_{k+1|k}, \hat{\mathbf{y}}_{k+1|k}^{(1)}) \ldots \mathsf{E}(\hat{\mathbf{y}}_{k+1|k}, \hat{\mathbf{x}}_{k+1|k}, \hat{\mathbf{y}}_{k+1|k}^{(2n)})\right) \ \mathbf{R}^{1/2}\right] \\ ② \ \mathsf{cholupdate}\left(\hat{\mathbf{S}}_{\mathbf{yy},k+1|k}, \mathsf{E}(\hat{\mathbf{y}}_{k+1|k}, \hat{\mathbf{x}}_{k+1|k}, \hat{\mathbf{y}}_{k+1|k}^{(0)}), W_{(c)}^{(0)}\right) \end{cases}$$

❺ Predict the cross-covariance of the invariant estimation errors:

$$\hat{\mathbf{P}}_{\mathbf{xy},k+1|k} = \begin{cases} \sum_{i=0}^{2n} W_{(c)}^{(i)} \eta(\boldsymbol{\mathcal{X}}_{k+1|k}^{(i)}, \hat{\mathbf{x}}_{k+1|k}) \mathsf{E}^T(\hat{\mathbf{y}}_{k+1|k}, \boldsymbol{\mathcal{X}}_{k+1|k}^{(i)}, \hat{\mathbf{y}}_{k+1|k}^{(i)}) \\ \sum_{i=0}^{2n} W_{(c)}^{(i)} \eta(\hat{\mathbf{x}}_{k+1|k}, \boldsymbol{\mathcal{X}}_{k+1|k}^{(i)}) \mathsf{E}^T(\hat{\mathbf{y}}_{k+1|k}, \hat{\mathbf{x}}_{k+1|k}, \hat{\mathbf{y}}_{k+1|k}^{(i)}) \end{cases}$$

► Correction/filtering step:

❶ Compute the gain terms: $\mathbf{K}_{k+1} = (\hat{\mathbf{P}}_{\mathbf{xy},k+1|k} / \hat{\mathbf{S}}_{\mathbf{yy},k+1|k}^T) / \hat{\mathbf{S}}_{\mathbf{yy},k+1|k}$

❷ Compute the new estimated state:

$$\hat{\mathbf{x}}_{k+1|k+1} = \hat{\mathbf{x}}_{k+1|k} + \sum_{i=1}^{n} \mathbf{K}_{k+1}^{(i)} \cdot \mathsf{E}(\hat{\mathbf{y}}_{k+1|k}, \hat{\mathbf{x}}_{k+1|k}, \mathbf{z}_{k+1}) \cdot w(\hat{\mathbf{x}}_{k+1|k})$$

❸ Update the matrix $\hat{\mathbf{S}}_{\mathbf{xx}}$:

$$\hat{\mathbf{S}}_{\mathbf{xx},k+1|k+1} = \mathsf{cholupdate}\left(\hat{\mathbf{S}}_{\mathbf{xx},k+1|k+1}, \mathbf{K}_{k+1}\hat{\mathbf{S}}_{\mathbf{yy},k+1|k}, -1\right)$$

4.8. Second reformulation of unscented Kalman filtering in an invariant setting: the π-IUKF algorithm

4.8.1. *Previous work on the IEKF algorithm*

To develop IEKF, S. Bonnabel and P. Rouchon established a formulation of classical EKF that incorporated ideas from invariant observer theory; the IEKF algorithm finds the linearized matrices that describe the dynamics of the invariant state estimation error. The covariance, which is the unknown of the Riccati equation, is redefined as the covariance matrix of the invariant errors. The equations of "standard" EKF with continuous dynamics are recalled below:

$$\dot{\hat{\mathbf{x}}} = f(\hat{\mathbf{x}}, \mathbf{u}) + \mathbf{K}(t)(\mathbf{y}_m - h(\hat{\mathbf{x}}, \mathbf{u})) \quad [4.31]$$

$$\mathbf{K}(t) = \mathbf{P}(t)\mathbf{C}^T(t)(\mathbf{N}\mathbf{N}^T)^{-1} \quad [4.32]$$

$$\dot{\mathbf{P}}(t) = \mathbf{A}(t)\mathbf{P}(t) + \mathbf{P}(t)\mathbf{A}^T(t) - \mathbf{P}(t)\mathbf{C}^T(t)(\mathbf{N}\mathbf{N}^T)^{-1}\mathbf{C}(t)\mathbf{P}(t) + \mathbf{M}\mathbf{M}^T \quad [4.33]$$

To compute the correction gain terms for the invariant estimator by EKF, we start by linearizing the invariant state estimation error $\eta(\mathbf{x}, \hat{\mathbf{x}}) = \varphi_{\mathbf{x}^{-1}}(\hat{\mathbf{x}}) - \varphi_{\mathbf{x}^{-1}}(\mathbf{x})$ to second order with the objective of identifying the matrices $\mathbf{A}$, $\mathbf{C}$, $\mathbf{N}$ and $\mathbf{M}$. Once these matrices are known, we can then deduce the gain matrix $\mathbf{K}$ from equations [4.31] to [4.33], which gives us the dynamics of the linearized error $\Delta\dot{\eta} = (\mathbf{A} - \mathbf{K}\mathbf{C})\Delta\eta$. This ultimately allows us to rewrite the IEKF equations as follows:

$$\dot{\hat{\mathbf{x}}} = f(\hat{\mathbf{x}}, \mathbf{u}) + \overbrace{DL_{\hat{\mathbf{x}}}}^{\substack{\text{invariant frame} \\ \text{at the point } \hat{\mathbf{x}} \sim w(\hat{\mathbf{x}})}} \cdot \mathbf{K}(t) \cdot \overbrace{(\rho_{\hat{\mathbf{x}}^{-1}}(\mathbf{z}) - \rho_{\hat{\mathbf{x}}^{-1}}(h(\hat{\mathbf{x}}, \mathbf{u})))}^{\mathrm{E}} \quad [4.34]$$

$$\mathbf{K}(t) = \mathbf{P}(t)\mathbf{C}^T(t)(\mathbf{N}\mathbf{N}^T)^{-1} \quad [4.35]$$

$$\dot{\mathbf{P}}(t) = \mathbf{A}(t)\mathbf{P}(t) + \mathbf{P}(t)\mathbf{A}^T(t) \quad [4.36]$$

$$-\mathbf{P}(t)\mathbf{C}^T(t)(\mathbf{N}\mathbf{N}^T)^{-1}\mathbf{C}(t)\mathbf{P}(t) + \mathbf{M}\mathbf{M}^T \quad [4.37]$$

In equation [4.34], the notation $DL_{\hat{\mathbf{x}}}$ indicates a left invariance of the Lie group G, but we could also consider a right invariance $DR_{\hat{\mathbf{x}}}$ (see Appendix).

Note that in order to construct an invariant observer on a Lie group, we must assume that the group action is free[5] over the state space, and that it has the same dimension as $\mathcal{X}$. As a result, there is only one single orbit; we say that the group action is transitive. The invariant observers established above for the AHRS and INS problems both satisfy this hypothesis.

Suppose that an invariant state estimation error η satisfying $\dot{\eta} = Dg_{ut}(\eta)$ is known, where Dg_{ut} is a vector field. Since G is a differentiable manifold, we can identify a piece of the Lie algebra of G with a neighborhood U of the group identity element e via the exponential mapping. Thus, any sufficiently small error η about e can be identified with some element ξ of the Lie algebra, in the sense that $\eta_t = \exp(\epsilon\xi_t)$, where $\epsilon \in \mathbb{R}$ satisfies $\epsilon << 1$. Up to second order in ϵ, the linearized equation of the invariant state estimation error η_t in the tangent space at e may be written as:

$$
\begin{aligned}
\frac{d}{dt}\xi &= [\xi, f(e, \hat{I})] - \frac{\partial f}{\partial u}(e, \hat{I})\frac{\partial \psi}{\partial g}(e, \hat{I})\xi \\
&\quad - \sum_{i=1}^{n}\left(\frac{\partial \mathbf{K}_i}{\partial \mathsf{E}}(\hat{I}, 0)\frac{\partial h}{\partial \mathbf{x}}(e, \hat{I})\xi\right) w_i,
\end{aligned}
\tag{4.38}
$$

where [,] is the Lie bracket. The matrices $\mathbf{A}(t)$ and $\mathbf{C}(t)$ can be identified as follows using equation [4.38]:

$$
\begin{aligned}
&\mathbf{A}(\hat{I}) : \xi \mapsto [\xi, f(e, \hat{I})] - \frac{\partial f}{\partial u}(e, \hat{I})\frac{\partial \psi}{\partial h}(e, \hat{I})\xi, \\
&\mathbf{C}(\hat{I}) = \frac{\partial h}{\partial x}(e, \hat{I}), \quad M = M(e), \quad N = N(e).
\end{aligned}
\tag{4.39}
$$

Other authors [BAR 14] have shown that approximating equation [4.38] rewritten as $\dot{\xi} = Dg_{ut}(\xi, e)$ by a first-order Taylor expansion yields results that are representative of the global behavior of the dynamics of the invariant state estimation error. In particular, the following theorem was established:

THEOREM 4.1.– Consider the invariant state estimation error η_t^i, defined with either left or right-invariant dynamics, $i = L$ or $i = R$. Define $\xi_0^i \in \mathbb{R}^{dim}(g)$

5 The full-rank hypothesis.

such that $\eta_0^i = \exp(\xi_0)$. If ξ_t^i satisfies a linear differential equation of the following form, for $t > 0$:

$$\frac{d}{dt}\xi_t^i = A_t^i \xi_t^i, \text{ where } A_t^i := Dg_{ut}^i(e),$$

then:

$$\forall t \geq 0,\ \eta_t^i = \exp(\xi_t^i). \qquad [4.40]$$

A full proof of this theorem can be found in [BAR 14]. This is an extremely important result, since it guarantees that the prediction step performed by any Kalman filter based on an invariant state error[6] – and thus implicitly based on an exponential chart of the geometric space – will preserve the true evolution of the estimated state error (in terms of its covariance).

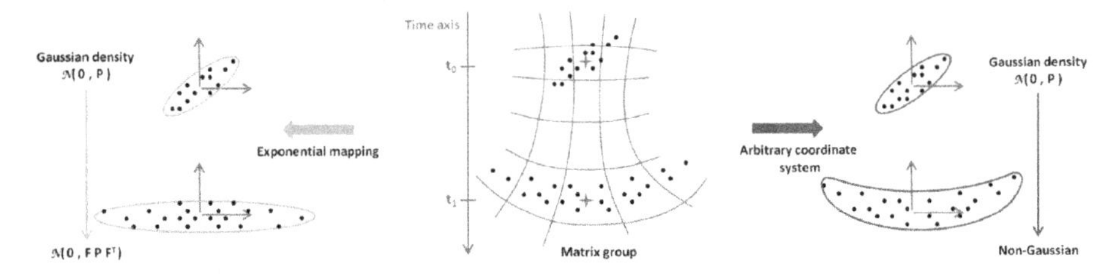

Figure 4.22. *Illustration of the log-linearity properties of the propagation error (reproduced from [BAR 14]). For a color version of this figure, see www.iste.co.uk/condomines/kalman.zip*

Figure 4.22 sketches an outline of the prediction step of a discrete nonlinear Kalman filter. This illustration shows the case where the state ranges over a manifold (such as the group $SO(\mathbb{R}^n)$). In the middle of the figure, the distribution of a Gaussian random variable $\hat{\mathbf{x}}_{t_0}$ defined by the fixed covariance matrix $\mathbf{P}$, represented graphically as a set of black points, is undergoing a transformation, carried by the tangent space via an evolution equation (as before, known as a "flow"). In the case where an arbitrary coordinate system is used to represent $\hat{\mathbf{x}}_{t_0}$, the transformation carried by the tangent space at time t_1 does not preserve the Gaussian distribution of the initial state, due to the nonlinearity of the flow, as can be very clearly seen on

6 Either $\eta = x^{-1}\hat{x}$ for left-invariant dynamics or $\eta = \hat{x}x^{-1}$ for right-invariant dynamics.

the right-hand side. But with a system of local coordinates that perfectly characterizes the tangent space of the manifold, using exponentials to construct a solution of the invariant state estimation error on the Lie group, the distribution of the cloud of points remains Gaussian and the covariance matrix at time t_1 becomes $\mathbf{F}_{t_1}\mathbf{P}\mathbf{F}_{t_1}^T$ (discrete equivalent of equation [4.37]).

4.8.2. *Formulation of the π-IUKF algorithm on a Lie group*

This section proposes another variant of the classical UKF algorithm presented in Chapter 2 for dynamic systems with symmetries. As noted in the introduction, one advantage of the UKF algorithm is that it computes an approximate solution to the optimal nonlinear problem of estimating the state of a dynamic system in discrete time, without requiring any linearization of the differential equations. The "sigma-points approach" (SPA) is typically used to approximate the random Gaussian process $x_{k|k}$ over time by applying the following nonlinear transformations:

$$\begin{cases} \mathbf{x}_k \sim \mathcal{N}(\hat{\mathbf{x}}_{k|k}, \mathbf{P}_{k|k}) \quad \text{such that} \quad \mathbf{x}_{k+1} = f(\mathbf{x}_k, k) + \mathbf{w}_k \\ \mathbf{y}_k = h(\mathbf{x}_k, k) + \mathbf{v}_k \end{cases} \qquad [4.41]$$

where $\mathbf{x} \in \mathbb{R}^n$, $\mathbf{y} \in \mathbb{R}^p$ and $\mathbf{w}$, $\mathbf{v}$ are Gaussian white noise terms. Recall that SPA aims to define a finite set of sigma points $\boldsymbol{\mathcal{X}}_{k|k}$ that capture the true mean $\hat{\mathbf{x}}_{k|k}$ and the true covariance $\mathbf{P}_{k|k}$ of the probability distribution characterizing $\mathbf{x}_{k|k}$. In the Gaussian case, these points need to accurately represent the first two moments of the random variable. The sigma points are then propagated through the nonlinear functions f and h from equation [4.41]. This gives a new cloud of points whose distribution differs from the original sigma points. The estimated mean $\hat{\mathbf{x}}_{k+1|k}$ and covariance $P_{k+1|k}$ of the transformed points are then computed from the statistical properties of the new set of points. The UKF algorithm can be viewed as a simple way of extending the SPA algorithm to compute the approximations needed to solve the estimation problem in the case where:

$$\mathbf{x}_{k+1} = f(\mathbf{x}_k, \mathbf{u}_k) + \mathbf{w}_k \qquad [4.42]$$

$$\mathbf{y}_k = h(\mathbf{x}_k, \mathbf{u}_k) + \mathbf{v}_k \qquad [4.43]$$

We shall assume that the evolution and observation noise terms ($\mathbf{w}$ and $\mathbf{v}$) are phrased in an invariant form. These noise terms can be viewed as exogenous inputs that leave the symmetries of the system intact. Accordingly, to avoid spoiling the symmetry, the noise must be integrated into the evolution and observation equations of the system as G-invariant/equivariant terms. This explains why some operations appear in the discrete differential equations of the estimator, such as right-multiplication by a quaternion $\cdot * q$.

To exploit the principles of UKF in invariant nonlinear settings, an additional condition on the observation equations is required, known as the *compatibility condition.* Specifically, a nonlinear function π is applied to the predicted output vector. It wouldn't make sense to use unmodified measurements in the modified variant of UKF proposed here, since an invariant output error is desired instead of a classical linear output error. Moreover, recall that our filtering equations project the various correction terms onto an invariant frame while preserving the symmetries of the system. The key idea of the π-IUKF algorithm is that the image of the output vector under the mapping π is related to the local diffeomorphism ρ_g and the invariant vector field. This invariant vector field gives a basis $TG(e)$ of the tangent space that is fixed when constructing the invariant coordinate system by the method of moving frames. In particular, this basis can be locally identified with the subgroup of quaternions of norm one by virtue of its construction. Additionally, the invariant output already appears in the invariant innovation term of any G-invariant estimator (see equation [4.5]). Consequently, the transformation π may be interpreted as a local mapping that constructs a vector of new predicted outputs whose action coincides with the group ρ at the identity element e. The components of the images can be expressed in terms of the natural basis $w_i^q(\hat{\mathbf{x}}_{k+1|k})$ of $TG(e)$ as follows:

$$\pi(h(\hat{\mathbf{x}}_{k+1|k}, \mathbf{u}_k)) = \sum_{i=1}^{3} \rho_{\gamma(\hat{\mathbf{x}}_{k+1|k})}(h(\hat{\mathbf{x}}_{k+1|k}, \mathbf{u}_k)) \cdot w_i^q(\hat{\mathbf{x}}_{k+1|k}) \cdot e_i. \qquad [4.44]$$

The compatibility condition guarantees that the output prediction step will preserve the symmetries of the system. It naturally follows that the gain terms of the SR-UKF algorithm presented in section 4.6 are functions of the known invariants $I(\hat{\mathbf{x}}, \mathbf{u})$, since $\hat{\mathbf{P}}_{\mathbf{yy},k+1|k}$ and $\hat{\mathbf{P}}_{\mathbf{xy},k+1|k}$ themselves depend on the result in equation [4.44] via this transformation. Note that the computation of the covariance matrices in the variant of the π-IUKF algorithm proposed here is still systematically based on a linear error. Therefore, π-IUKF uses almost exactly the same computation steps as classical UKF. The output prediction step is modified by introducing the transformation π, and a geometrically adapted correction phrased in terms of an invariant output error is used in the filtering equations, instead of a linear correction term based on a linear innovation term. Note that the covariance matrix of the state estimation error must be initialized with values from an invariant estimation error $\eta(\hat{x}_0, x_0)$. In practice, this tends not to matter, since $\hat{\mathbf{P}}_0$ is usually assumed to be $>> 1$. In summary, the stochastic differential equations describing the π-IUKF algorithm are established by modifying the model output prediction step (by introducing the transformation π) then adapting the filtering equations in the correction step.

PROPOSITION 4.1.– If a mapping π satisfying equation [4.44] can be constructed analytically, then a UKF technique can be used to calculate the correction gain terms in equation [4.5] by sampling the stochastic distributions of the estimated state and output errors while preserving the symmetries of the problem.

It follows that equation [4.44] can be written explicitly for the models $\mathcal{M}^+_{ahrs}$ and $\mathcal{M}^+_{ins}$ as follows:

$$\pi(h_{ahrs}(\hat{\mathbf{x}}_{k+1|k})) = \begin{pmatrix} \hat{a}_{s_{k+1|k}} \cdot \sum_{i=1}^{3} (A * \hat{q}_{k+1|k}) \cdot e_i \\ \hat{b}_{s_{k+1|k}} \cdot \sum_{i=1}^{3} (B * \hat{q}_{k+1|k}) \cdot e_i \end{pmatrix},$$

$$\pi(h_{ins}(\hat{\mathbf{x}}_{k+1|k})) = \begin{pmatrix} \sum_{i=1}^{3} (\hat{V}_{k+1} * \hat{q}_{k+1}) \cdot e_i \\ \sum_{i=1}^{3} (\hat{X}_{k+1} * \hat{q}_{k+1}) \cdot e_i \\ < \hat{X}_{k+1} * \hat{q}_{k+1}, e_3 > -\hat{h}_b \\ \sum_{i=1}^{3} (B * \hat{q}_{k+1}) \cdot e_i \end{pmatrix}. \quad [4.45]$$

Pseudocode for the π-IUKF algorithm is listed at the end of this chapter. Four configuration parameters need to be defined:

1) the scalar λ, which captures the moments of the distribution of the random variable $\hat{\mathbf{x}}_{k|k}$. The notation $(\hat{\mathbf{P}})^{1/2}_{k|k}$ denotes the square root matrix of the covariance matrix, i.e. $(\hat{\mathbf{P}})^{1/2}_{k|k} = \hat{\mathbf{S}}_{k|k}$ such that $\hat{\mathbf{P}}_{k|k} = \hat{\mathbf{S}}_{k|k}\hat{\mathbf{S}}^T_{k|k}$;

2) the weights $W^{(j)}_{(m)}$ and $W^{(j)}_{(c)}$ associated with the distributions of the mean and the covariance, satisfying $\sum_{j=0}^{2n} \mathcal{W}^{(j)}_{(m,c)} = 1$;

3) finally, $\mathbf{W}_k$ and $\mathbf{V}_k$, the estimated covariance matrices of the noise terms, used to counteract the effect of noise on the model.

Note that the "square root" form of the UKF method is listed here (SR-UKF). For convenience, the operation of adding an arbitrary vector $\mathbf{u}$ to each column of a matrix $\mathbf{M}$ is written $\mathbf{M} \pm \mathbf{u}$. The SR-UKF approach improves the computational efficiency and numerical stability of the algorithm. It is based on three techniques from linear algebra ➠ QR factorization (written qr$\{\cdot\}$); ➠ rank-one updating of the Cholesky factor (MATLAB command TM cholupdate); ➠ solving a linear system by a least-squares technique (MATLAB command TM \).

Pseudocode of the π-IUKF algorithm with additive noise terms:

– Initialization:

$$\hat{\mathbf{x}}_0 = \mathbb{E}[\mathbf{x}_0] \qquad S_0 = \text{chol}\Big\{\mathbb{E}[\eta(\hat{\mathbf{x}}_0, \mathbf{x}_0)\eta(\hat{\mathbf{x}}_0, \mathbf{x}_0)^T]\Big\}$$

– For $kT = 1, \ldots, \infty$:

- ❶ Compute the sigma points:

$$\boldsymbol{\chi}_{k|k} = [\quad \hat{\mathbf{x}}_{k|k} \quad \underbrace{\chi_{k|k}^{(1)} \cdots \chi_{k|k}^{(n)}}_{\hat{\mathbf{x}}_{k|k} + \sqrt{n+\lambda}\sqrt{P_{k|k}}} \quad \underbrace{\chi_{k|k}^{(n+1)} \cdots \chi_{k|k}^{(2n)}}_{\hat{\mathbf{x}}_{k|k} + \sqrt{n+\lambda}\sqrt{P_{k|k}}} \quad]$$

- ❷ Prediction equations:

$$\boldsymbol{\chi}_{k+1|k} = \left[\chi_{k+1|k}^{(0)} \cdots \chi_{k+1|k}^{(2n+1)}\right] = f(\boldsymbol{\chi}_{k|k}, \mathbf{u}_k)$$

$$\hat{\mathbf{x}}_{k+1|k} = \sum_{i=0}^{2n} \mathcal{W}_{(m)}^{(i)} \chi_{k+1|k}^{(i)}$$

$$S_{k+1|k} = \text{qr}\Big\{\Big[\sqrt{\mathcal{W}_{(c)}^{(1)}}\Big(\Big[\chi_{k+1|k}^{(1)} \cdots \chi_{k+1|k}^{(2n)}\Big] - \hat{\mathbf{x}}_{k+1|k}\Big) \quad \sqrt{\mathbf{W}_k}\Big]\Big\}$$

$$S_{k+1|k} = \text{cholupdate}\{S_{k+1|k}, \chi_{k+1|k}^{(0)} - \hat{\mathbf{x}}_{k+1|k}, \mathcal{W}_{(c)}^{(0)}\}$$

- ❸ Correction equations:

$$\hat{\mathbf{y}}_{k+1|k}^{i} = \left[y_{k+1|k}^{(0)} \cdots y_{k+1|k}^{(2n+1)}\right] = \pi(h(\boldsymbol{\chi}_{k+1|k}, u_k))$$

$$\hat{\mathbf{y}}_{k+1|k} = \sum_{i=0}^{2n} \mathcal{W}_{(m)}^{(i)} \mathbf{y}_{k+1|k}^{(i)}$$

$$S_{y,k+1|k} = \text{qr}\Big\{\Big[\sqrt{\mathcal{W}_{(c)}^{(1)}}\Big[\hat{\mathbf{y}}_{k+1|k}^{i}(:, 2:2n+1) - \hat{\mathbf{y}}_{k+1|k}\Big] \quad \sqrt{\mathbf{V}_{k+1}}\Big]\Big\}$$

$$S_{y,k+1|k} = \text{cholupdate}\{S_{y,k+1|k}, \mathbf{y}_{k+1|k}^{(0)} - \hat{\mathbf{y}}_{k+1|k}, \mathcal{W}_{(c)}^{(0)}\}$$

$$P_{xy,k+1|k} = \sum_{i=0}^{2n} \mathcal{W}_{(c)}^{(i)} \Big[\chi_{k+1|k}^{(i)} - \hat{\mathbf{x}}_{k+1|k}\Big]\Big[\mathbf{y}_{k+1|k}^{(i)} - \hat{\mathbf{y}}_{k+1|k}\Big]^T$$

- ❹ Compute the gain terms of the Kalman filter:

$$\bar{\mathbf{K}}_{k+1} = (P_{xy,k+1|k}/S_{z,k+1|k}^T)/S_{z,k+1|k}$$

$$\hat{\mathbf{x}}_{k+1|k+1} = \hat{\mathbf{x}}_{k+1|k} + \bar{\mathbf{K}}_{k+1}(\pi, E) \cdot E \cdot w_i^q(\hat{\mathbf{x}}_{k+1|k})$$

$$U = \bar{\mathbf{K}}_{k+1} S_{y,k+1|k}$$

$$S_{k+1|k+1} = \text{cholupdate}\Big\{S_{k+1|k}, U, -1\Big\}$$

5

Methodological Validation, Experiments and Results

This chapter presents a comprehensive set of results obtained using the various algorithmic methods presented in the previous chapters. Tests were performed on simulated data, then on real data. After evaluating the performance of the algorithms, we sought to study and compare their characteristic behavior, invariant and otherwise. Our results validate the overall approach of this book, namely the idea that a UKF-type technique might be able to adapt the generic framework of invariant observer theory to the specific problem of constructing a nonlinear estimate of the state of a dynamic system. The first part of this chapter compares the performance of the newly proposed π-IUKF algorithm and the square root variant of scaled UKF (SR-UKF) presented in section 4.4. This analysis, performed on simulated data, begins by considering the complete AHRS model, then is extended to the complete INS model. Both models incorporate descriptions of various imperfections in the sensors. The second part of this chapter considers experimental results for both models, constructed from real data by each of the three key algorithms presented throughout this book: SR-UKF, π-IUKF and the invariant observer algorithm. For the invariant observer, we present a methodology that can be used to easily configure the gain terms governing the dynamics of the algorithm. A suitable configuration can be found by constructing a knowledge base of *a priori* information about the behavior of each state variable as a function of the system gain terms. This can be done by analyzing the observer on a large set of simulated results derived from real data. The knowledge base is then aggregated and represented graphically using a technique inspired by fuzzy logic. The final part of this chapter is dedicated to a comparative study of the navigation performance of the three algorithms. An extremely accurate localization system known as the Differential Global Positioning System (DGPS) provides reference measurements of the position and velocity of the UAV, giving a basis for comparison to determine which algorithm constructs the best estimate. Each algorithm was coded in the MATLAB/SIMULINK programming environment. The embedded code of the invariant observer has already been incorporated into the avionics systems of the mini-UAVs used at the ENAC laboratory.

5.1. Validation with simulated data

In this section, we evaluate the estimation performance of the newly developed π-IUKF algorithm against the performance of the standard SR-UKF algorithm. Both algorithms are applied to the navigation models presented in sections 3.5.2 and 3.6.1 to process simulated data representing the aerial release of a payload suspended from a parachute (see section 4.5.2).

5.1.1. *Presentation of the data*

The new technologies presented in Chapter 2 have opened up a wide range of prospective applications for aerial release operations. This is one of the topics currently being studied by the ONERA[1] institute in France. In particular, ONERA is developing sports-like parachutes with an added guidance system that allows more accurate airdrops from an aircraft. An inertial measurement system composed of a triad of sensors (accelerometer, magnetometer and gyrometer) is embedded onto the payload, providing extremely accurate inertial data. These data offer the perfect opportunity for us to study the performance of the various estimation algorithms presented throughout this chapter. The payload carries a high-performance inertial sensor, as well as a GPS, which provides information about the velocity. ONERA provided us with a set of measurements describing the aerial release of a payload, recorded at a sample rate of 50 Hz, and arranged into a matrix of 12 noise-free and unbiased parameters characterizing the navigation of the payload. These parameters were as follows:

---

1 *Office National d'Etudes et de Recherches Aérospatiales*, National Office for Aerospace Study and Research.

1) angular velocity of the payload relative to three axes $\omega_m = (\omega_{mx}\ \omega_{my}\ \omega_{mz})^T$;

2) specific acceleration of the payload relative to three axes $a_m = (a_{mx}\ a_{my}\ a_{mz})^T$;

3) a local measurement of the Earth's magnetic field, expressed in the coordinate system of the payload, $y_B = (B_{mx}\ B_{my}\ B_{mz})^T$;

4) velocity of the payload, expressed in ground coordinates $(V_x\ V_y\ V_z)^T$.

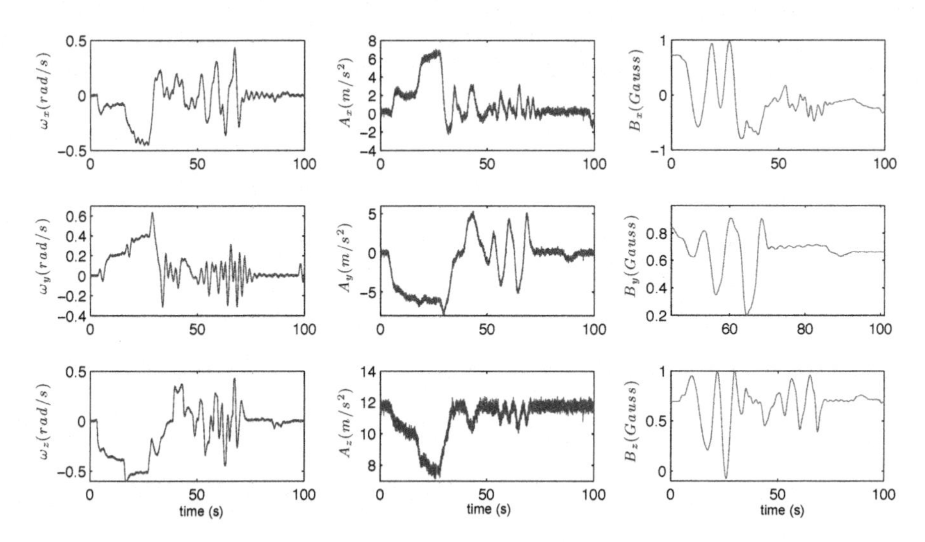

Figure 5.1. *Simulated sensor measurements: angular velocity - total acceleration - Earth's magnetic field. For a color version of this figure, see www.iste.co.uk/condomines/kalman.zip*

We begin by validating the models $\mathcal{M}_{ahrs}$ and $\mathcal{M}_{ins}$ presented in Chapter 3 using the input data provided by ONERA. By integrating the equations of the two models with respect to time, we can accurately reconstruct the state representing the attitude of the payload, and hence its navigation, in full. This reconstruction gives a reference against which we can evaluate the performance of each estimation algorithm. Gaussian white noise terms are added to these reference measurements to represent the imperfections in each sensor, allowing us to study the performance of each algorithm under more realistic conditions. A selection of noisy measurements is shown in Figure 5.1. The noise was generated with specific standard deviation values by a mini-UAV simulation environment called Aviones

[RAN 07]. The numerical characteristics of the imperfections are listed in Table 5.1. The perturbation terms are directly introduced as follows into the models $\mathcal{M}^+_{ahrs}$ and $\mathcal{M}^+_{ins}$ for the π-IUKF algorithm:

$$\mathcal{M}^+_{ahrs} \begin{cases} q_{k+1} = \frac{1}{2} q_k * (\omega_{m_k} - \omega_{b_k}) + W_{q_k} w_{q_k} * q_k \\ w_{b_{k+1}} = q_k{}^{-1} * W_{w_k} w_{w_k} * q_k \\ a_{s_{k+1}} = a_{s_k} W_{a_k} w_{a_k} \\ b_{s_{k+1}} = b_{s_k} W_{b_k} w_{b_k} \\ \begin{pmatrix} z_{A_{k+1}} \\ z_{B_{k+1}} \end{pmatrix} = \begin{pmatrix} a_{s_{k+1}} q_{k+1}^{-1} * (A + V_{A_{k+1}} v_{A_{k+1}}) * q_{k+1} \\ b_{s_{k+1}} q_{k+1}^{-1} * (B + V_{B_{k+1}} v_{B_{k+1}}) * q_{k+1} \end{pmatrix} \end{cases} \quad [5.1]$$

$$\mathcal{M}^+_{ins} \begin{cases} q_{k+1} = \frac{1}{2} q_k * (\omega_{m_k} - \omega_{b_k}) + W_{q_k} w_{q_k} * q_k \\ V_{k+1} = A + \frac{1}{a_{s_k}} q_k * a_{m_k} * q_k^{-1} + W_{V_k} w_{V_k} \\ X_{k+1} = V_k + W_{X_k} w_{X_k} \\ \omega_{b_{k+1}} = q_k{}^{-1} * W_{w_k} w_{w_k} * q_k \\ a_{s_{k+1}} = a_{s_k} W_{a_k} w_{a_k} \\ h_{b_{k+1}} = h_{b_k} W_{b_k} w_{b_k} \\ \begin{pmatrix} z_{V_{k+1}} \\ z_{X_{k+1}} \\ z_{h_{k+1}} \\ z_{B_{k+1}} \end{pmatrix} = \begin{pmatrix} V_{k+1} + V_{V_{k+1}} v_{V_{k+1}} \\ X_{k+1} + V_{X_{k+1}} v_{X_{k+1}} \\ < X_{h_{k+1}} \mid e_3 > - h_{b_{k+1}} + V_{h_{k+1}} v_{h_{k+1}} \\ q_{k+1}^{-1} * (B + V_{B_{k+1}} v_{B_{k+1}}) * q_{k+1} \end{pmatrix} \end{cases} \quad [5.2]$$

In equations 5.1 and 5.2, the terms $W_{q_k}, W_{V_k}, W_{X_k}, W_{w_k}, W_{a_k}$ and W_{b_k} (respectively, V_{A_k}, V_{B_k}, V_{V_k} and V_{X_k}) are the weights associated with the covariance matrices of the noise in the state (respectively, in the measurements). In our case, the configuration used to generate the simulation results was defined by choosing the elements of the covariance matrix V_k of the measurement noise to reflect the true noise values of the sensors, which are listed in Table 5.1. Similarly, the covariance matrix of the noise in the state is defined by $W_{q_k} = 0.001 I_3, W_{V_k} = 0.01 I_3, W_{X_k} = 0.01 I_3, W_{w_k} = 0.001 I_3, W_{a_k} = 0.001 I_3$ and $W_{b_k} = 0.001 I_3$. The invariant term

$\lambda(1 - \|\hat{q}_k\|^2)\hat{q}_k$ is also added to the kinematic formula of the prediction models of each algorithm as follows (setting $\lambda = 1$ in our case):

$$\dot{\hat{q}}_{k+1} = \frac{1}{2}\hat{q}_k * (\omega_{m_k} - \hat{\omega}_{b_k}) + K_{q_k}E * \hat{q}_k + \lambda(1 - \|\hat{q}_k\|^2)\hat{q}_k.$$

This invariant term is designed to compensate any numerical errors in the quaternions. Finally, for consistency with the models $\mathcal{M}^+_{ahrs}$ and $\mathcal{M}^+_{ins}$, an additive bias vector $\omega_b = (0.1\,\text{rad/s}\,0.05\,\text{rad/s}\,0.02\,\text{rad/s})$ is introduced into the measurement ω_m, and two scale factors are added, denoted a_s, b_s for the AHRS and a_s, h_b for the INS.

Parameter	Value	Unit
$\sigma_{gyro,x}$	0.005	rad/s
$\sigma_{gyro,y}$	0.005	rad/s
$\sigma_{gyro,z}$	0.005	rad/s
$\sigma_{acc,x}$	0.005	m/s
$\sigma_{acc,y}$	0.005	m/s
$\sigma_{acc,z}$	0.005	m/s
$\sigma_{mag,x}$	0.005	Gauss
$\sigma_{mag,y}$	0.005	Gauss
$\sigma_{mag,z}$	0.005	Gauss
$\sigma_{GPS,px}$	1	m
$\sigma_{GPS,py}$	1	m
$\sigma_{GPS,pz}$	1	m
$\sigma_{GPS,vx}$	0.5	m/s
$\sigma_{GPS,vy}$	0.5	m/s
$\sigma_{GPS,vz}$	0.5	m/s

Table 5.1. *Characteristic parameters of the measurement noise in each sensor*

Each of the algorithms presented in Chapter 4 needs to be initialized with some initial state $\hat{\mathbf{x}}_{0|0}$. We chose to compute our initial state from the first available measurements. To do this, we need to evaluate the value of each component of the quaternion and the linear velocity at the initial state. The initial quaternion was calculated from the accelerometer and magnetometer measurements as follows:

$$\begin{cases} \varphi & = \arctan2(-a_{my}, -a_{mz}) \\ \theta & = \arcsin\left(\dfrac{-a_{mx}}{\|a_{mx}\|}\right) \\ \psi & = \arctan2((B_z \cdot \sin(\varphi) - B_y \cdot \cos(\varphi), \cdots \\ & B_x \cdot \cos(\theta) + (B_y \cdot \sin(\varphi) + B_z \cos(\varphi)) \cdot \sin(\theta)) \end{cases}$$

These values can then be converted into a quaternion as follows:

$$\begin{cases} q_0 = \cos(\psi/2) \cdot \cos(\theta/2) \cdot \cos(\varphi/2) + \sin(\psi/2) \cdot \sin(\theta/2) \cdot \sin(\psi/2) \\ q_1 = \cos(\psi/2) \cdot \cos(\theta/2) \cdot \sin(\varphi/2) + \sin(\psi/2) \cdot \sin(\theta/2) \cdot \cos(\psi/2) \\ q_2 = \sin(\psi/2) \cdot \cos(\theta/2) \cdot \sin(\varphi/2) + \cos(\psi/2) \cdot \sin(\theta/2) \cdot \cos(\varphi/2) \\ q_3 = \sin(\psi/2) \cdot \cos(\theta/2) \cdot \cos(\varphi/2) + \cos(\psi/2) \cdot \sin(\theta/2) \cdot \sin(\varphi/2) \end{cases}$$

Finally, the GPS measurements at the first sampled point were used to define the initial linear velocity of the payload.

5.1.2. *Performance study of the two UKF algorithms on the AHRS model*

5.1.2.1. *Validation without any error in the initial state*

We begin by testing the algorithms in the case where the initial state $\hat{\mathbf{x}}_{0|0} = (0.99\ 0\ \ -0.0103\ 0\ 0\ \ 0\ 0\ 1\ 1)^T$ is known. To make the simulation more realistic, the bias terms ω_b defined above are added to the state, as well as the scale factors $a_s = 1.2$ and $b_s = 0.9$. Finally, for both algorithms, the covariance of the estimation error is defined in terms of the standard deviations $\sqrt{P_0} = diag(0.0001 I_4\ 0.002 I_3\ 0.001 I_2)$, where I_n is the $n \times n$ identity matrix. To present the results more intuitively, the quaternions were converted into Euler angles. Figure 5.2 shows the estimated attitude angles (φ, θ, ψ) produced by each algorithm. Already in our first set of results, we can see that the estimation error of the π-IUKF algorithm is minimal, even after introducing noise into the evolution and observation models. Note also that the estimation error in the heading is slightly higher for π-IUKF than for UKF between 20 and 30 s. However, this does not significantly affect the attitude of the micro-UAV, since the amplitude of the error remains small in either case.

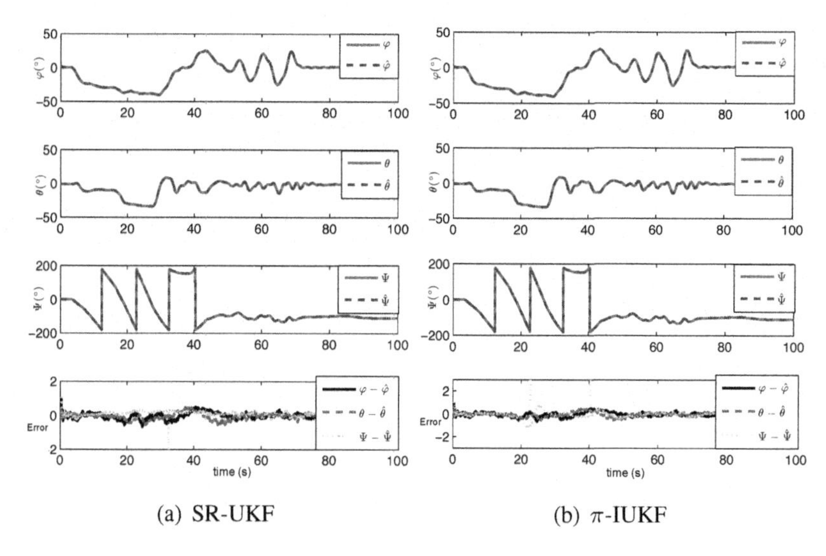

(a) SR-UKF (b) π-IUKF

Figure 5.2. *Estimated attitude angles (φ, θ, ψ). For a color version of this figure, see www.iste.co.uk/condomines/kalman.zip*

The next task is to check the validity of the AHRS model against the simulation throughout the entire duration of the airdrop. Figure 5.3 plots the norm of the acceleration $\dot{V}$, showing that Hypothesis 3.4 from section 3.5 (the assumption that $\dot{V} = 0$) is not perfectly satisfied. The norm of the acceleration fluctuates strongly both upward and downward in the period between 20 and 35 s. This deceleration phase occurs because the parachute is deployed a few minutes after the airdrop. Figure 5.4 shows a few small inconsistent perturbations in the biases and scale factors between 20 and 40 s. During this phase, the acceleration is at its highest, and hence the hypothesis $\dot{V} = 0$ does not hold. Note that this phase of strong acceleration influenced the bias terms because they were estimated while the system was still changing, which would not have been the case in practice. Figure 5.4(a) shows that there is still a slight error in the final estimates. The largest error is in $\hat{a}_s$. Nevertheless, for both algorithms, the estimated biases converge rapidly to values that are representative of the biases introduced into the measurements.

5.1.2.2. *Performance of the algorithms*

The estimation errors in the attitude and bias terms can be used to evaluate the performance of each algorithm more precisely. Figure 5.6 shows the mean of the error and the theoretical and statistical values of the standard deviation in each case. To calculate the standard deviation, we chose to use the criterion defined in the following equation:

$$\text{CRITERION:} J(k) = \sqrt{\frac{\sum_{i=k}^{n+k} (\mathbf{x}_i - \hat{\mathbf{x}}_i)^2}{n}} \qquad [5.3]$$

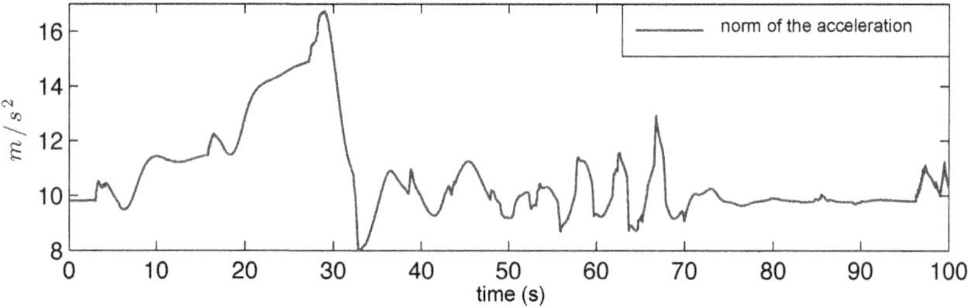

Figure 5.3. *Norm of the acceleration y_A (the assumption $\dot{V} \neq 0$ does not hold locally)*

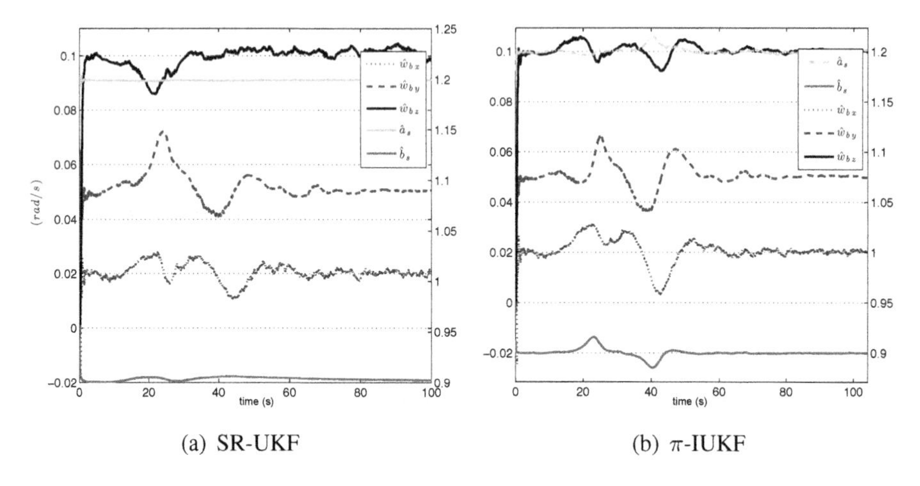

Figure 5.4. *The bias and scale factors are locally affected by the fact that $\dot{V} \neq 0$. For a color version of this figure, see www.iste.co.uk/condomines/kalman.zip*

This criterion is based on the Root Mean Square Deviation (RMSD). It allows us to quantitatively evaluate the difference between the quaternions and reference biases predicted by the theoretical model and the estimated values constructed by each method.

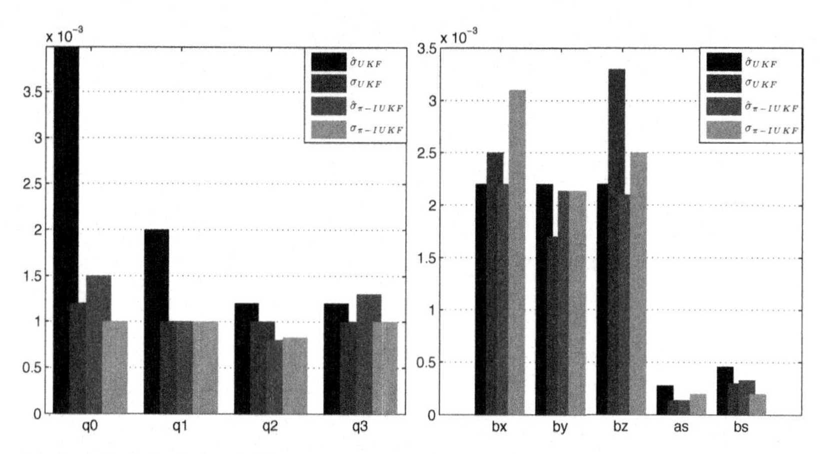

(a) Standard deviation of the quaternions (b) Standard deviation of the bias terms

Figure 5.5. *Comparison of the theoretical/measured standard deviations for the SR-UKF and π-IUKF algorithms. For a color version of this figure, see www.iste.co.uk/condomines/kalman.zip*

The results shown in Figure 5.6 reveal a very clear distinction between the convergence of the covariance matrices of SR-UKF and π-IUKF. Figure 5.6(a) shows that the covariance matrix of the first quaternion converges to an incorrect value. This is because the SR-UKF algorithm does not structurally preserve unit norms, unlike π-IUKF. By comparison with the convergence properties of the invariant observer presented in Chapter 4, the covariance matrices of π-IUKF do indeed converge to constant values in every case.

Finally, observe that the estimation errors converge to zero, which shows that π-IUKF is capable of reconstructing the attitude of the reference model. Figure 5.5 compares the estimated covariance errors constructed by both algorithms ($\hat{\sigma}_{UKF}$ and $\hat{\sigma}_{\pi-IUKF}$) against those calculated using the criterion J (σ_{UKF} and $\sigma_{\pi-IUKF}$). The performance of both algorithms is essentially equivalent. Still, it is worth noting that the estimated standard deviation values proposed by π-IUKF are closer to the true values than those computed by SR-UKF.

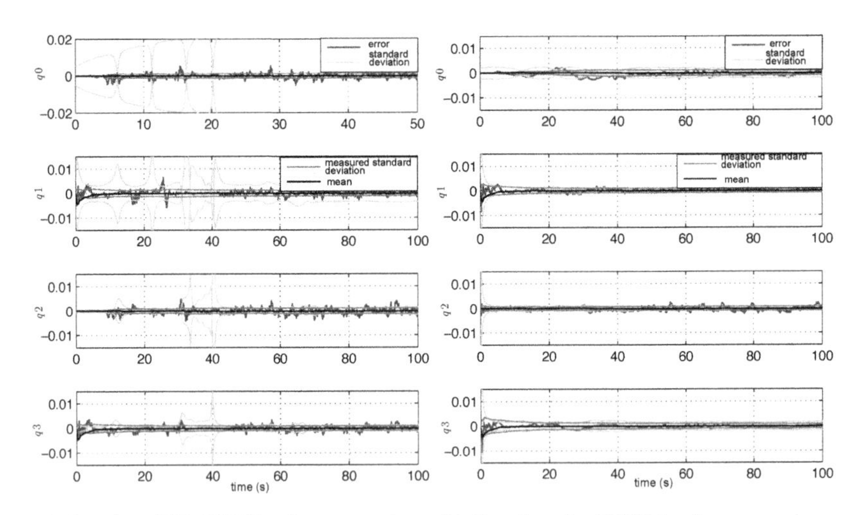

(a) Results of SR-UKF for the quaternions (b) Results of π-IUKF for the quaternions

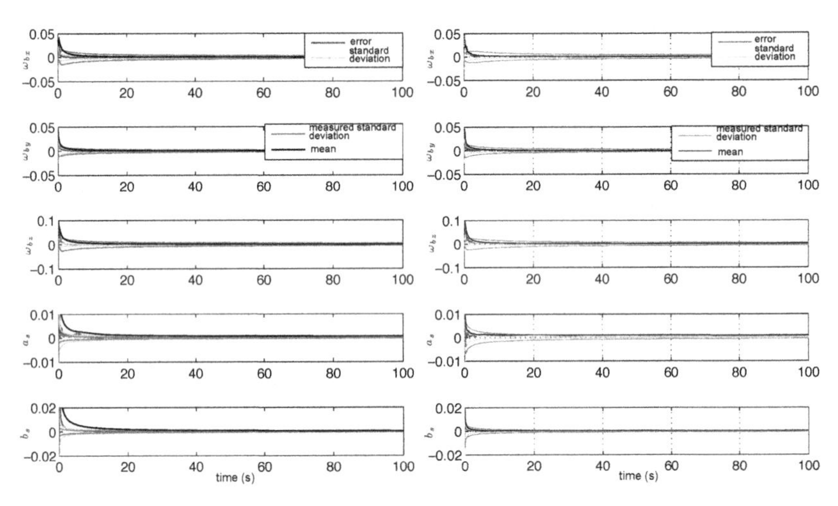

(c) Results of SR-UKF for the bias terms (d) Results of π-IUKF for the bias terms

Figure 5.6. *Results of the SR-UKF and π-IUKF algorithms. For a color version of this figure, see www.iste.co.uk/condomines/kalman.zip*

5.1.2.3. *Introducing an initial error*

We now introduce an error into the initial state vector $\hat{\mathbf{x}}_{0|0}$ by rotating through an angle of 120° away from the true orientation of the system. The quaternions of the initial state are now:

	True system	Initial state
q_0	0.99	$\cos(\pi/3)$
q_1	0	$\sin(\pi/3)/\sqrt{(3)}$
q_2	-0.0103	$-\sin(\pi/3)/\sqrt{(3)}$
q_3	0	$\sin(\pi/3)/\sqrt{(3)}$

For both algorithms, the covariance of the estimation error is defined with the standard deviation values $\sqrt{P_0} = diag(0.01I_4\ 0.002I_3\ 0.001I_2)$. The objective of this section is to demonstrate that π-IUKF is capable of estimating the attitude even when the initial conditions deviate significantly from the true state of the system. The theoretical Euler angles (φ, θ, ψ) are plotted as a function of time in Figure 5.7, showing that the π-IUKF algorithm converges more rapidly than its counterpart. Although both SR-UKF and π-IUKF were initialized far away from the true values, the time frame required for their state estimates to converge is sufficiently short to be satisfactory in the context of our application (computing the attitude of a mini-UAV).

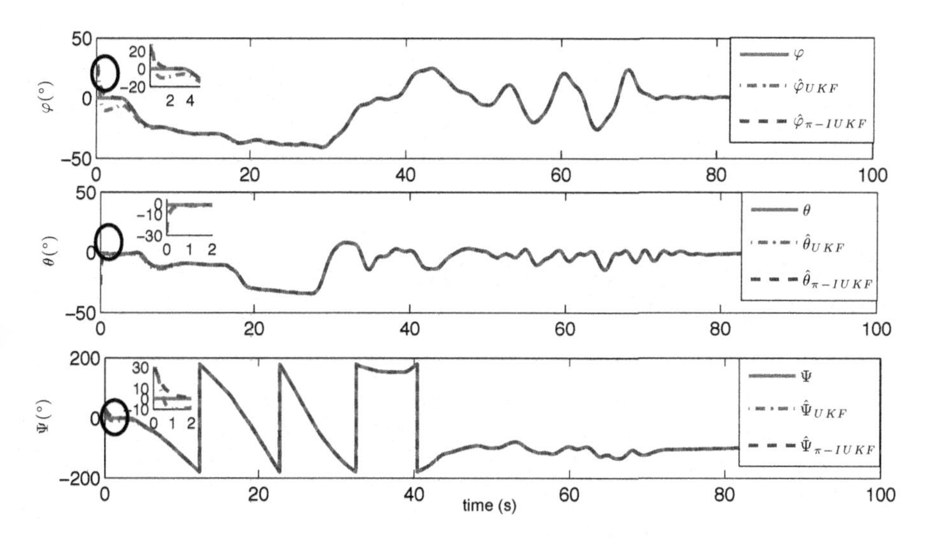

Figure 5.7. *Convergence of the SR-UKF and π-IUKF. For a color version of this figure, see www.iste.co.uk/condomines/kalman.zip*

5.1.3. *Invariance properties of π-IUKF on the AHRS model*

Although we have established that π-IUKF behaves well in terms of reconstructing the state, we still need to verify that it has the same invariance properties as the IEKF algorithm [BON 09b]. The π-IUKF algorithm can be viewed as a left-invariant observer that uses the gain terms of a UKF filter to compute the error. We therefore need to check that the estimated quaternion $\hat{q}(t)$ and hence the convergence (and the gain terms) are now independent of the trajectory, despite the presence of nonlinear dynamics. To demonstrate the first key invariance property satisfied by the convergence behavior of the state, consider again the configuration from the previous section, but now choosing the initial conditions so that the estimated and true orientations differ by an angle of $60°$. This gives the following quaternions:

	TRUE SYSTEM	INITIAL STATE
q_0	0.99	$\cos(\pi/6)$
q_1	0	$\sin(\pi/6)/\sqrt{(3)}$
q_2	-0.0103	$-\sin(\pi/6)/\sqrt{(3)}$
q_3	0	$\sin(\pi/6)/\sqrt{(3)}$

Figure 5.8 shows the convergence of $\hat{q}$ to q for the π-IUKF algorithm. The figure also shows the norm of the linear state estimation error between q and $\hat{q}$ computed by SR-UKF, as well as the invariant state estimation error computed by π-IUKF. For the latter error, observe that the norm of the elements of the group $\eta = \hat{q} \cdot q^{-1}$ converge to 1 as the norm of the estimation error $||\hat{q} \cdot q^{-1} - 1||$ tends to zero.

Next, consider another set of initial conditions for the theoretical system is chosen so that the norm of the state error is equal to the norm of the error computed above for the initial state. For example, we can choose the following values:

	TRUE SYSTEM	INITIAL STATE
q_0	0.5	$\cos(\pi/6)$
q_1	0.5	$\sin(\pi/6)/\sqrt{(3)}$
q_2	0.5	$-\sin(\pi/6)/\sqrt{(3)}$
q_3	0.5	$\sin(\pi/6)/\sqrt{(3)}$

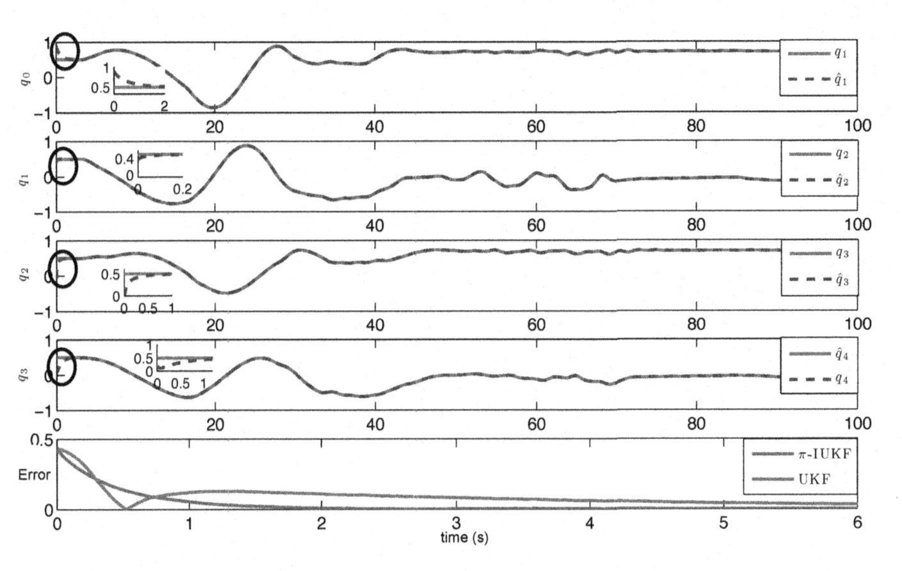

Figure 5.8. *Estimated attitude $\hat{q}$ (blue dotted lines) constructed by π-IUKF, as well as the linear estimation error (SR-UKF) and the invariant estimation error (π-IUKF). For a color version of this figure, see www.iste.co.uk/condomines/kalman.zip*

These values satisfy $\|\hat{q}(0) \cdot q^{-1}(0) - 1\| = 1$ and $\|q(0) - \hat{q}(0)\| = 0.5176$. With this new set of initial conditions, the invariant state error behaves exactly as before, even though the trajectory is now different (see Figure 5.9). The same is not true for the classical state error calculated by the SR-UKF algorithm.

The second invariance property of the π-IUKF filter concerns the gain terms $\mathbf{K}$. Returning to the same configuration used earlier to estimate the Euler angles (see Figure 5.2), we now consider the evolution of the gain terms calculated by UKF and π-IUKF as the system travels along its trajectory. Figure 5.10 compares the gain matrices constructed by UKF to the values computed by π-IUKF. The correction coefficients of π-IUKF become constant after just a few seconds ($t < 5$ s) and remain constant over the entire trajectory. By contrast, the coefficients of SR-UKF behave irregularly, and the value of the SR-UKF estimation error remains far away from its equilibrium point. After $t > 75$ s, the trajectories of the system settle into a permanent (nearly constant) regime, at which point the coefficients of the SR-UKF gain terms also stabilize.

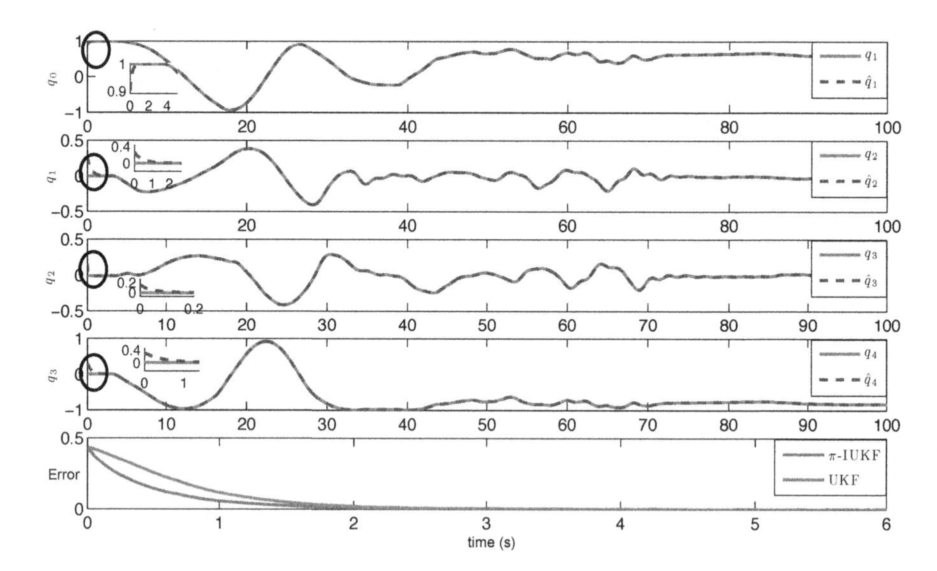

Figure 5.9. *Estimated attitude $\hat{q}$ (blue dotted lines) constructed by π-IUKF, as well as the linear estimation error (SR-UKF) and the invariant estimation error (π-IUKF). The only difference relative to Figure 5.8 is the choice of initial conditions q_0. For a color version of this figure, see www.iste.co.uk/condomines/kalman.zip*

The presented simulations show that the convergence of π-IUKF has a very global character. The π-IUKF algorithm succeeded in computing numerical values for the correction gain terms that give the filter the desired dynamics while ensuring that the invariant state estimation error converges on a large domain.

5.1.4. *Performance study of the two UKF algorithms on the INS model*

5.1.4.1. *Validation with known initial conditions*

We now present the results obtained by applying each of the two UKF algorithms to the INS model. Figure 5.11 shows the estimates of the attitude angles (φ, θ, ψ) computed by UKF and π-IUKF, respectively. In the period from 60 to 80 s, the errors in θ and ψ are larger for UKF than for π-IUKF. This shows that π-IUKF is capable of reconstructing the attitude with a small estimation error even when significant noise is introduced into the evolution and observation models (see Table 5.1).

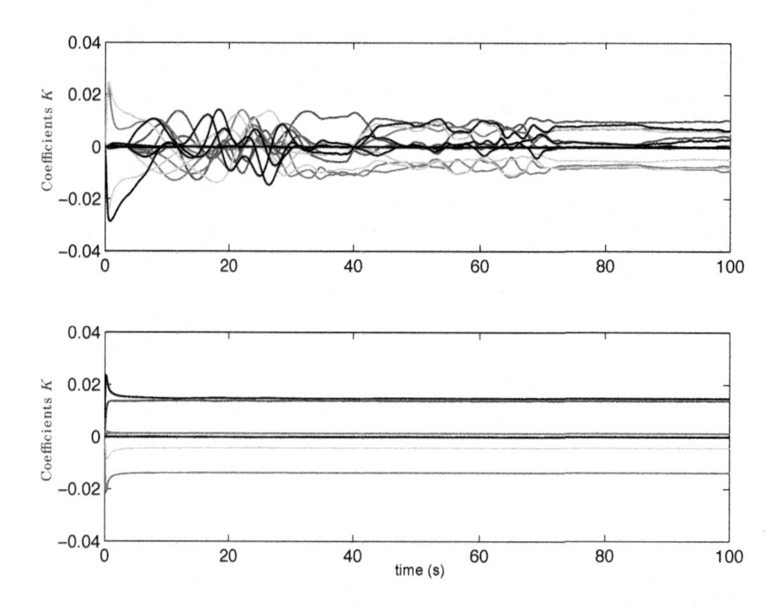

Figure 5.10. *Convergence of the correction gain terms K for SR-UKF (top) and π-IUKF (bottom). For a color version of this figure, see www.iste.co.uk/condomines/kalman.zip*

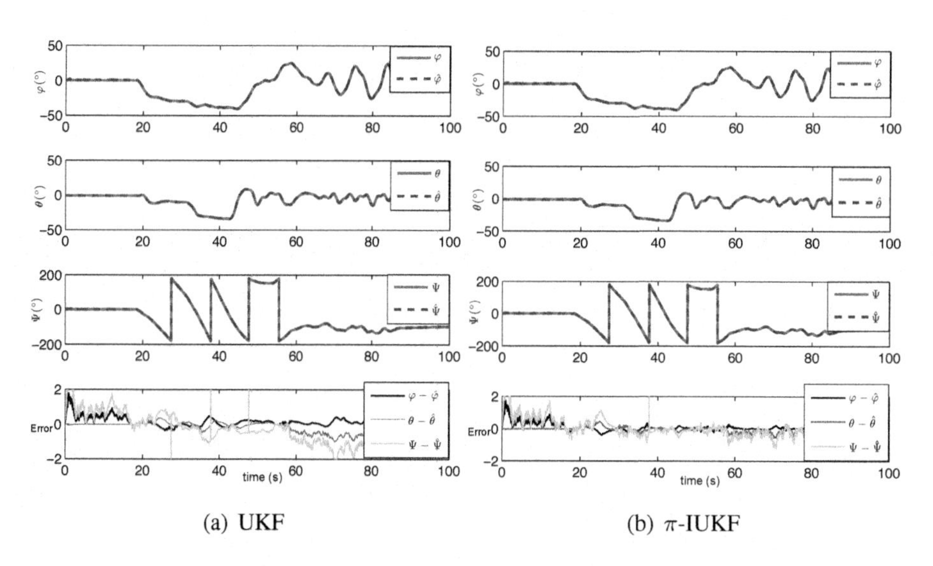

Figure 5.11. *Estimated attitude (φ, θ, ψ). For a color version of this figure, see www.iste.co.uk/condomines/kalman.zip*

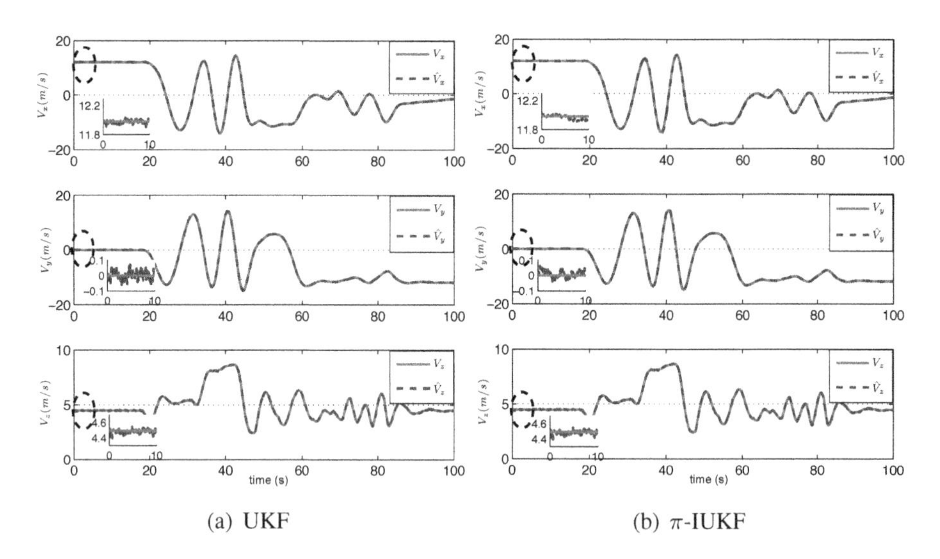

Figure 5.12. *Estimated attitude (V_x, V_y, V_z). For a color version of this figure, see www.iste.co.uk/condomines/kalman.zip*

Figures 5.12 and 5.13 show the velocity and position values computed by each algorithm. First, observe that the amplitudes of the velocity data ($\pm 20\,\mathrm{m/s}$) and position data ($\pm 500\,\mathrm{m}$) are consistent with the flight profile of a fixed-wing mini-UAV. If we magnify the first 10 s of velocity data, the figure seems to suggest that the estimation performance of π-IUKF is slightly better. This would need to be confirmed by conducting a more in-depth performance study. Additionally, if we magnify the first half-second, we see that the y-component of the position converges more slowly for UKF than for π-IUKF. Both algorithms correctly estimated the positions. These two results show that the π-IUKF algorithm is able to properly reconstruct the position and velocity of the payload despite significant noise in the evolution and observation models of the INS.

5.1.4.2. *Performance of the two algorithms*

The next step is to evaluate the performance of each algorithm more precisely by considering the estimation errors in the attitude, velocity and position, as well as the bias in the gyrometers, accelerometers and barometers. As before, we define the covariance of the estimation error to have standard deviations $\sqrt{P_0} = diag(0.0001I_4,\ 0.002I_3,\ 0.001I_3,\ 0.01I_3,\ 0.05I_2)$ for both algorithms. Figures 5.15 and 5.16 present the results obtained, showing the mean error and the theoretical and statistical values of the standard deviation in each case, calculated from equation 5.3. Figures 5.14(a) and 5.14(b) compare the values of the covariance errors estimated by each algorithm against the criterion J from equation 5.3.

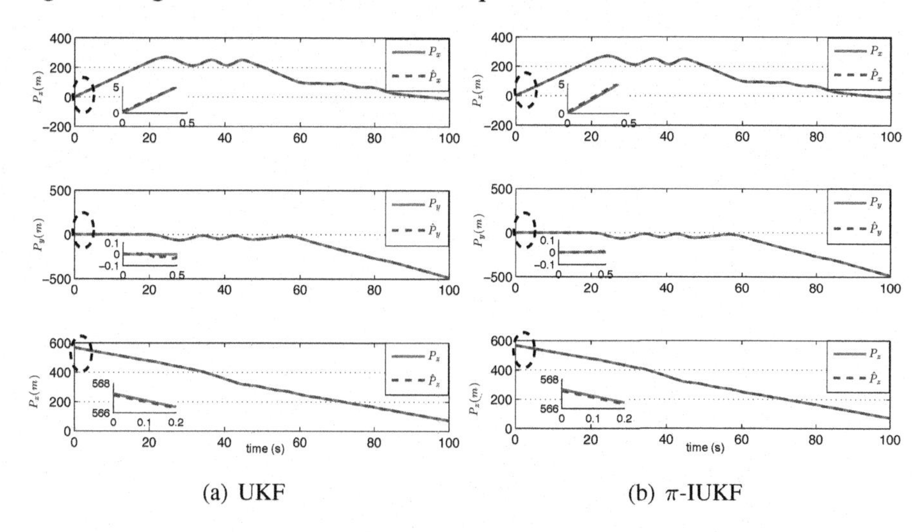

Figure 5.13. *Estimated position (P_x, P_y, P_z). For a color version of this figure, see www.iste.co.uk/condomines/kalman.zip*

The results exhibit very similar behavior to the above AHRS model. In particular, the unit norm of the quaternion is not structurally preserved by the SR-UKF algorithm, which leads to an incorrect value for the covariance of the q_0-component. Note that there are slight variations in the covariances calculated by π-IUKF, which may be due to the invariant state error no longer being autonomous, unlike the AHRS model. Figures 5.14(a) and 5.14(b) show that the standard deviation estimated by π-IUKF is comparatively closer to the true standard deviation than the SR-UKF estimate. Overall, both algorithms perform similarly and produce very good estimates of the attitude, biases, position and velocity.

5.1.4.3. *Introducing an initial error into the state vector*

We now introduce an initial error into the state vector $\hat{\mathbf{x}}_{0|0}$, as we did earlier for the AHRS model. To study the convergence of the estimated state under more realistic conditions, we assume that the system is immobile for around 10 s. As well as a rotation of 120° relative to the true orientation of the system, we assume that the velocity of the initial state is far away from its true value. Finally, in practice, the flight coordinates can differ significantly from the integrated coordinates in the algorithm due to terrain mapping inaccuracies. To reflect this source of error, which primarily affects the X- and Y- coordinates, we choose position values that are a few meters away from the true state. For both algorithms, the covariance of the estimation error is defined with standard deviations $\sqrt{P_0} = diag(0.01I_4,\ 4,\ 0.8,\ 0.8,\ 15,\ 15,\ 0.1,\ 0.002I_3,\ 0.001I_2)$. These initial conditions are summarized in the following graphs:

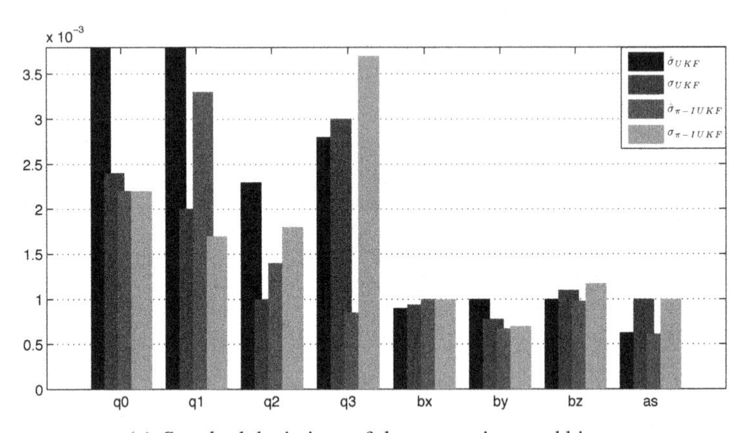

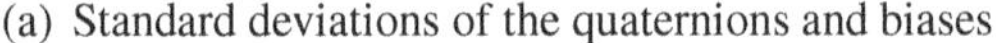
(a) Standard deviations of the quaternions and biases

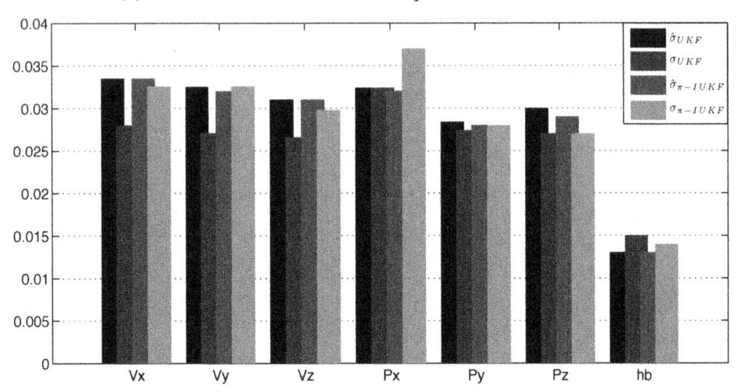

(b) Standard deviations of the velocities and positions

Figure 5.14. *Comparison of the theoretical/measured standard deviations for the SR-UKF and π-IUKF algorithms. For a color version of this figure, see www.iste.co.uk/condomines/kalman.zip*

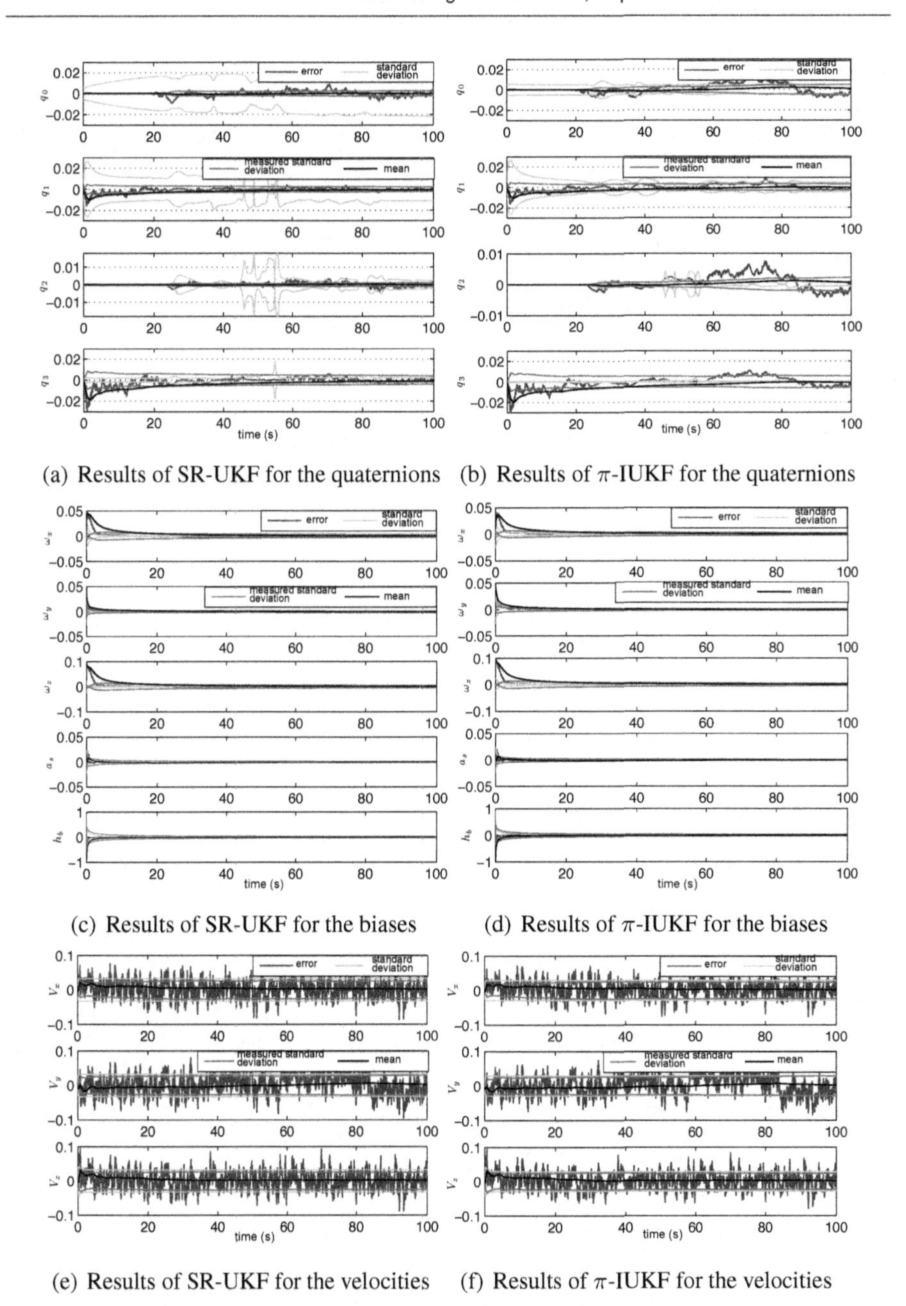

(a) Results of SR-UKF for the quaternions

(b) Results of π-IUKF for the quaternions

(c) Results of SR-UKF for the biases

(d) Results of π-IUKF for the biases

(e) Results of SR-UKF for the velocities

(f) Results of π-IUKF for the velocities

Figure 5.15. *Results of π-IUKF and SR-UKF. For a color version of this figure, see www.iste.co.uk/condomines/kalman.zip*

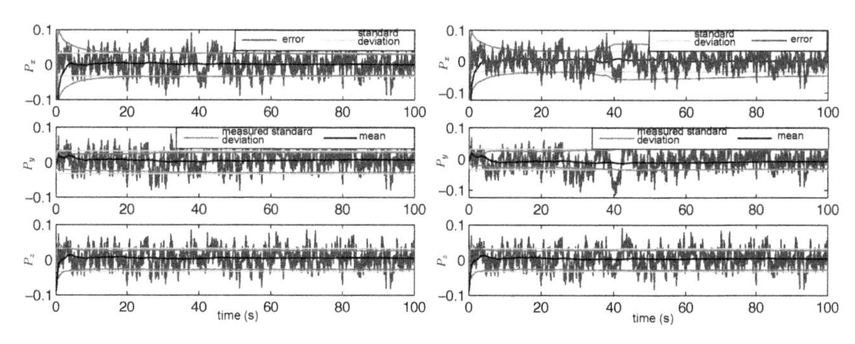

(a) Results of SR-UKF for the positions (b) Results of π-IUKF for the positions

Figure 5.16. *Comparison of the results of SR-UKF and π-IUKF. For a color version of this figure, see www.iste.co.uk/condomines/kalman.zip*

	TRUE SYSTEM	INITIAL STATE
q_0	0.99	$\cos(\pi/3)$
q_1	0	$\sin(\pi/3)/\sqrt{(3)}$
q_2	-0.0103	$-\sin(\pi/3)/\sqrt{(3)}$
q_3	0	$\sin(\pi/3)/\sqrt{(3)}$
V_x	14.78	2
V_y	0	2
V_z	4.75	2
P_x	0	50
P_y	0	50
P_z	567	567

Figure 5.17(a) shows that π-IUKF achieved much better convergence for the attitude angle φ. The π-IUKF algorithm required less than 5 s to converge, whereas the SR-UKF took over 15 s, even though both algorithms use the same covariance matrix for the estimation error. The convergence of the other attitude angles was similar for both algorithms. The velocities and positions are shown in Figures 5.17(b) and 5.18. Again, π-IUKF achieved noticeably better convergence. Nevertheless, both algorithms converged within a relatively short time frame, and both algorithms succeeded in correctly reconstructing the state, despite the initial values being far away from the "true" state.

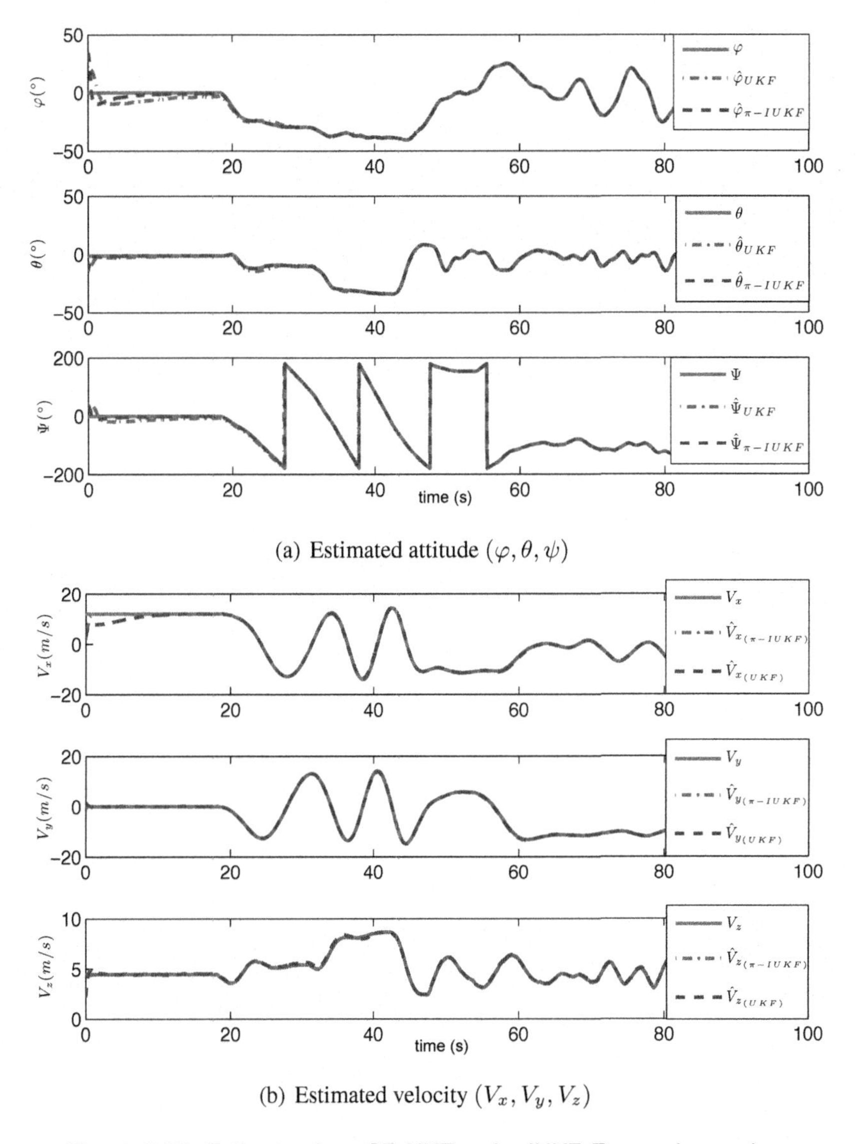

(a) Estimated attitude (φ, θ, ψ)

(b) Estimated velocity (V_x, V_y, V_z)

Figure 5.17. *Estimates from SR-UKF and π-IUKF. For a color version of this figure, see www.iste.co.uk/condomines/kalman.zip*

5.1.5. *Invariance properties of π-IUKF on the INS model*

The invariance properties of the INS model are naturally going to be narrower than those established earlier for the AHRS model. In section 5.3.1, the invariance properties were deduced from the convergence behavior of the gain terms over time on each trajectory of the system. Similarly, we again consider the behavior of the gain terms calculated by SR-UKF and π-IUKF over time for the INS model. For each state variable, Figure 5.19 compares the values of the SR-UKF gain matrices against the corresponding values constructed by π-IUKF. The quaternions no longer satisfy all of the properties established earlier for the AHRS; nevertheless, from the gain values, we can see that the domain of convergence of the invariant state estimation error was larger for π-IUKF by virtue of its invariance properties. Figures 5.19(b) and 5.19(c) clearly show that the invariant state error is no longer autonomous.

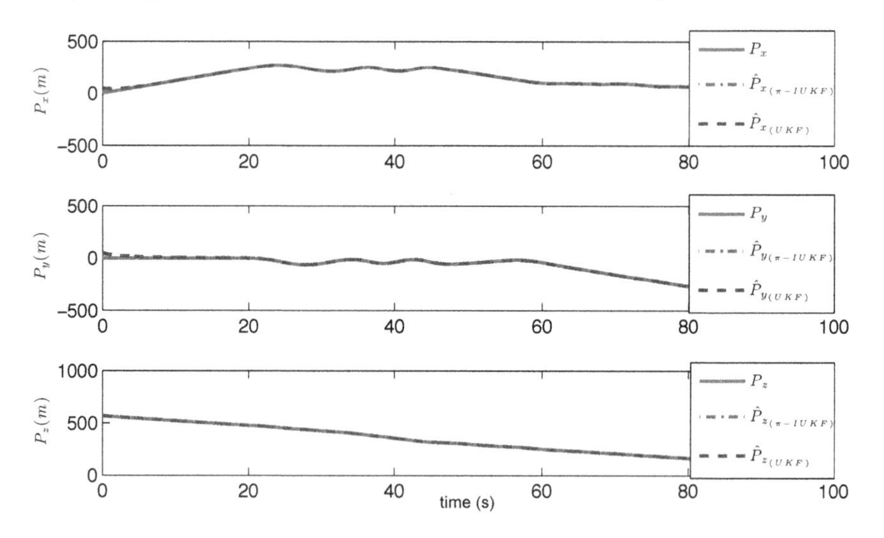

Figure 5.18. *Estimated position (P_x, P_y, P_z). For a color version of this figure, see www.iste.co.uk/condomines/kalman.zip*

The AHRS and INS simulation results did not reveal any significant issues with the convergence of the covariance of the estimation error for the π-IUKF algorithm, nor any significant differences in performance between π-IUKF and SR-UKF. Nevertheless, we observed in sections 5.1.3 and 5.1.5 that π-IUKF has an appreciable edge over UKF in terms of convergence. We judged the various estimates established from these simulations to be very encouraging, prompting us to consider experimental data.

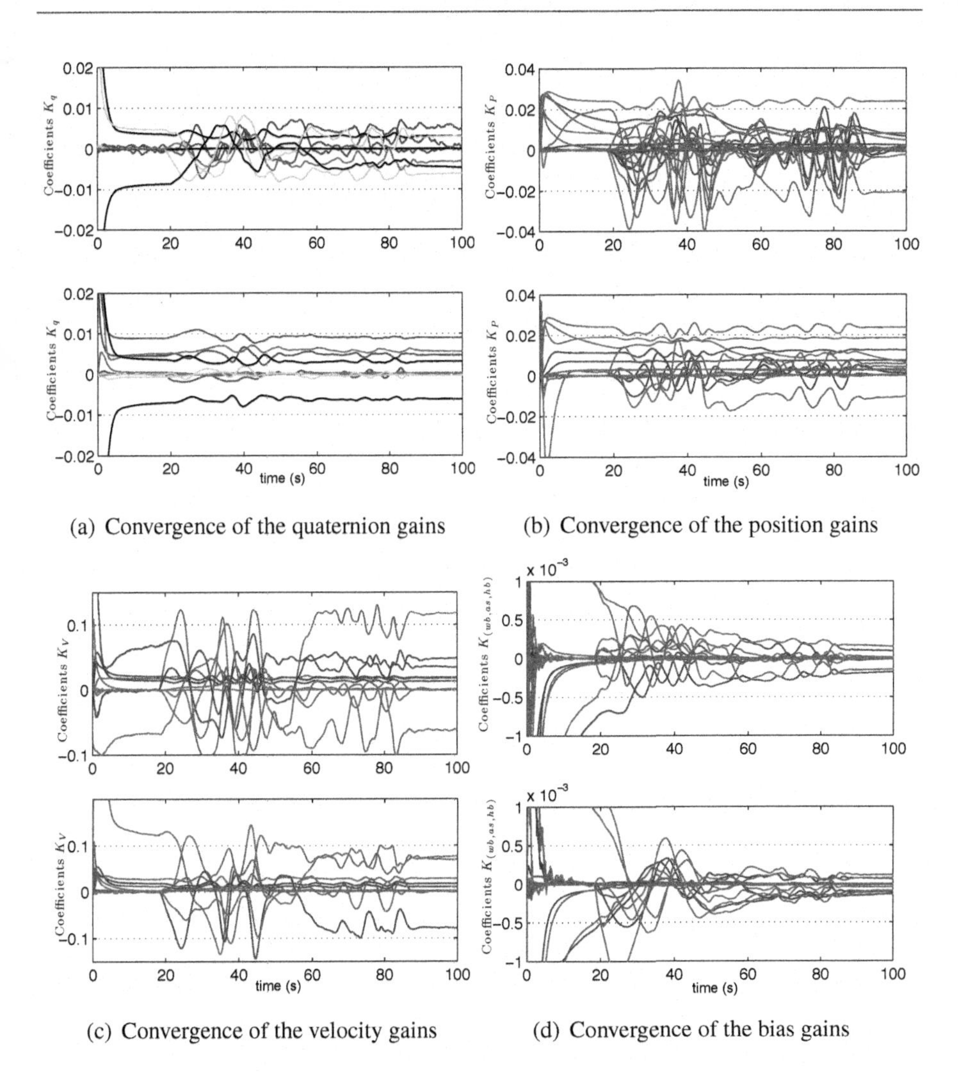

(a) Convergence of the quaternion gains

(b) Convergence of the position gains

(c) Convergence of the velocity gains

(d) Convergence of the bias gains

Figure 5.19. *Convergence of the gain coefficients K constructed by SR-UKF (top) and π-IUKF (bottom). For a color version of this figure, see www.iste.co.uk/condomines/kalman.zip*

5.2. Validation with real data

This section experimentally validates the estimation performance of the π-IUKF algorithm for mini-UAV navigation by considering both a complete AHRS model and a complete INS model. The first instrument of our experimental set-up is an Inertial Measurement Unit (IMU) designed at

ENAC that measures the acceleration, the magnetic field, the altitude and the angular velocity as outputs. Also in the case of IMU, we used two geolocalization antennas: a low-accuracy Ublox 6H GPS equipped with a passive antenna used to directly control the navigation of the mini-UAV (velocity and position), and a Ublox 6T GPS equipped with a high-accuracy active Novatel antenna used to generate reference measurements of the velocity and position of the microdrone by postprocessing the codes and phases (see Chapter 1) with the RTKLIB software package.

After describing the global performance of π-IUKF on the AHRS model, we present a preliminary experimental test performed to investigate the performance of the invariant observer on the INS model. We propose a straightforward method for configuring the gain terms that determine the dynamics of the observer based on a technique inspired by fuzzy logic. We then validate our algorithms for a micro-UAV in flight by performing a second experiment. The second and final part of this chapter is dedicated to an analysis of the performance of the position and velocity estimates by comparing the results of each algorithm to the Differential Global Positioning System (DGPS) reference measurements.

5.2.1. *Tools and experimental setup*

To experimentally validate the performance of the three algorithms presented in Chapter 4, we consider real inertial, magnetic and altimetric measurements taken directly from the mini-UAV. The data were recorded by the Paparazzi[2] autopilot chip known as "Apogee", as shown in Figure 5.20. The Apogee chip features an IMU, a magnetometer and a barometer. Based on the study of low-cost MEMS sensors presented in Chapter 2, we selected the sensors listed in Figure 5.20. The Novatel antenna shown in Figure 5.21 can only be mounted on a medium-sized drone, since it weighs around 150 g.

As well as various outputs for controlling the aircraft, the Apogee chip has a miniSD card that allows real-time data acquisition of the numerical outputs, acceleration and angular velocity, sampled at a frequency of 125 Hz for the inertial measurements and 10 Hz for the two others. The data acquisition

2 See www.wiki.paparazziuav.org.

module is placed in the center of the aircraft and connected to a GPS that measures the position and velocity at a frequency of 5 Hz.

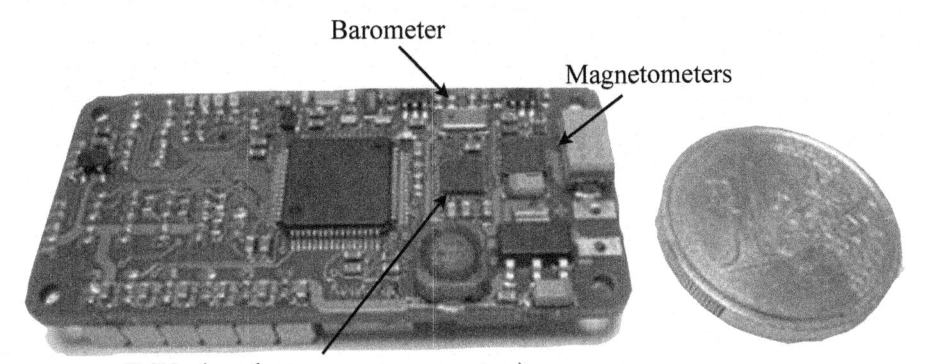

Description	Weight (g)	Price ($)	Size (mm)	Frequency (Hz)
Magnetometer HMC5883L	18	6.95	3, 3, 0.9	10
IMU MPU-6050	18	12.95	4, 4, 0.9	125
Barometer MPL3115A2	15	4.95	5, 3, 1.2	10

Figure 5.20. *Apogee autopilot chip. For a color version of this figure, see www.iste.co.uk/condomines/kalman.zip*

Figure 5.21. *Novatel antenna*

The results presented below were prepared in the simulation environment MATLAB/SIMULINK from the real data recorded by the chip. Figure 5.22 shows the Simulink model; the physical measurements are calibrated, then sent to the filter with output state vector x.

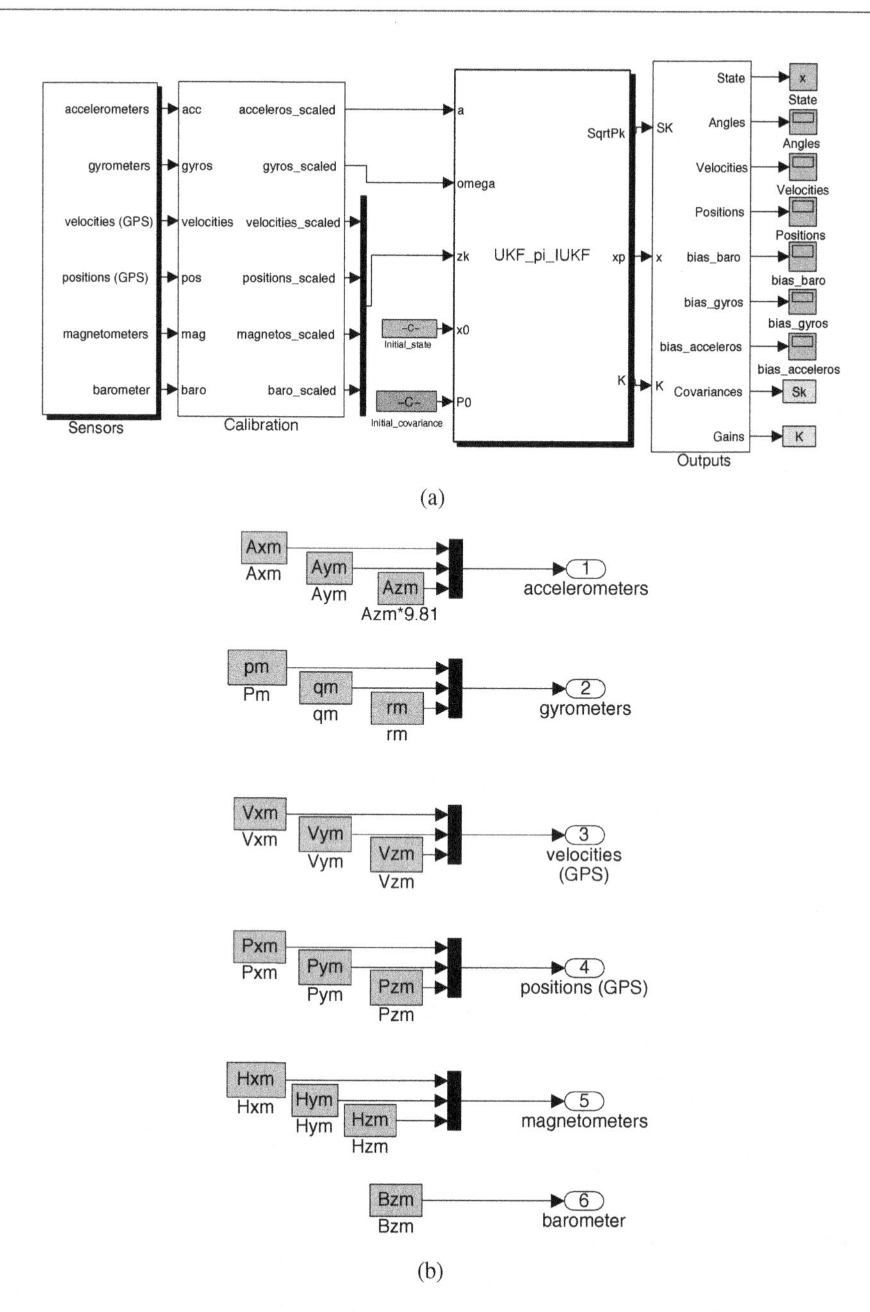

Figure 5.22. *Simulink diagram showing the acquisition and calibration of the measurements as well as an overview of the inputs/outputs of the SR-UKF and π-IUKF filters. For a color version of this figure, see www.iste.co.uk/condomines/kalman.zip*

5.2.2. *Preliminary identification of imperfections*

Following on from the presentation of the Allan variance method in Chapter 3, this section aims to identify the imperfections in the inertial sensors in order to establish the best possible representation of their flaws. As described in section 5.1.1, additive noise terms were introduced into the evolution and observation equations of the AHRS and INS models. We now need to accurately determine the spectral densities $\mathbf{Q}$ and $\mathbf{R}$ associated with the noise in the states and outputs of each model. To find the values of these parameters, we recorded 20 h of static data from the accelerometers and gyrometers at a frequency of 100 Hz in order to deduce the Allan variance graphically. For the other sensors (barometer, magnetometer and GPS), we calculated the standard deviations from recordings that were shorter but sufficiently long for the Gaussian distribution of the noise to be visible.

The first step is to identify the covariance of the measurement noise R classically for each sensor using the criterion stated in equation 5.3. The corresponding values are listed in Tables 5.2 and 5.3. In each case, the nominal spectral density $\mathbf{R}$ (in μg or $^{\circ}/s/\sqrt{(Hz)}$ specified by the manufacturer either overestimated or underestimated the true values. The standard deviations of the measurements from the magnetometers and the barometer were 0.0031 Gauss and 0.8 m. In Figure 5.24, we observe that the probability density of each noise term is either Gaussian or almost Gaussian, and so the hypotheses of the Kalman filter stated in Chapter 2 are met.

	Static bias	σ	R (measured)	R (manufacturer)	Q
a_{xm}	0.142	0.0319	325	400	$3.3e^{-4}$
a_{ym}	-0.3	0.0985	1000	400	$3.2e^{-4}$
a_{zm}	0.19	0.049	500	400	$5.2e^{-4}$
ω_{xm}	-1.55	0.0825	0.0083	0.0224	$1.4e^{-4}$
ω_{ym}	-1.13	0.1673	0.0167	0.0224	$2.66e^{-4}$
ω_{zm}	-1.7	0.2214	0.0221	0.0224	$3.16e^{-4}$

Table 5.2. *Static variance and bias of the readings from the Apogee chip*

The second step is to analyze the Allan variance in order to determine the values of $\mathbf{Q}$ associated with the spectral density of the noise in the state for each inertial sensor. The Allan variance equations presented in Chapter 3 can be used to generate a graph to read off each type of variance present in the noisy signal. For example, by considering the characteristic point $\tau = 1$, we can read $\sigma_{ARW} = \mathbf{Q}$ directly from the curve by its envelope of slope $-1/2$. Figures 5.23(a) and 5.23(b) show the time graph of each Allan variance associated with the three axes of the gyroscope and the accelerometer. The values shown are taken into account by the configurations of the SR-UKF and π-IUKF algorithms throughout the rest of the study on real data.

GPS	σ
X	0.88 m
Y	1.45 m
Z	20.29 m
V_X	0.248 m/s
V_Y	0.146 m/s
V_Z	0.280 m/s

Table 5.3. *Variance of the GPS measurements*

As stated above, the analysis of the Allan variance enables us to characterize any imperfections in the inertial sensors (such as noise) over the entire frequency spectrum, both at short-term and long-term scales. However, this technique is not enough to identify high-frequency noise (such as vibrations) during flight maneuvers. In the literature, the most common approach to handling high-frequency noise is to perform wavelet denoising [ALP 92] to improve the signal-to-noise ratio. However, this is too computationally expensive to be viable for our application. To study this problem, we performed a frequency analysis based on the Fourier spectrum of the signal, computed from the noise in the real data recorded by the gyrometers and accelerometers during each flight phase (waiting for take-off, take-off, cruise flight and landing). Figure 5.25 plots the angular velocity and acceleration measurements recorded on the miniSD card during flight. We defined four phases over the full period of 400 s: (1) waiting for take-off; (2) take-off; (3) cruise flight; (4) landing. Note that the noise in the accelerometer measurements a_{mx} is distinct for each of these four phases. Figure 5.26 demonstrates this analysis along one axis with a sampling frequency of 250 Hz. This allow us to determine a useful frequency band that excludes any vibrations caused by the aircraft. After analyzing the three axes of the accelerometers and gyrometers, we configured the internal low-pass filter of each inertial sensor to have a cut-off frequency of 20 Hz.

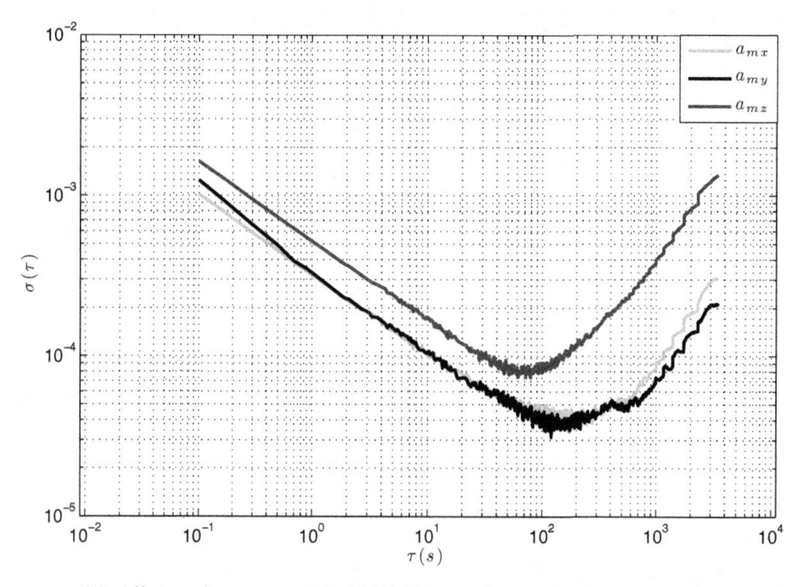

(a) Allan variance associated with the accelerometer measurements

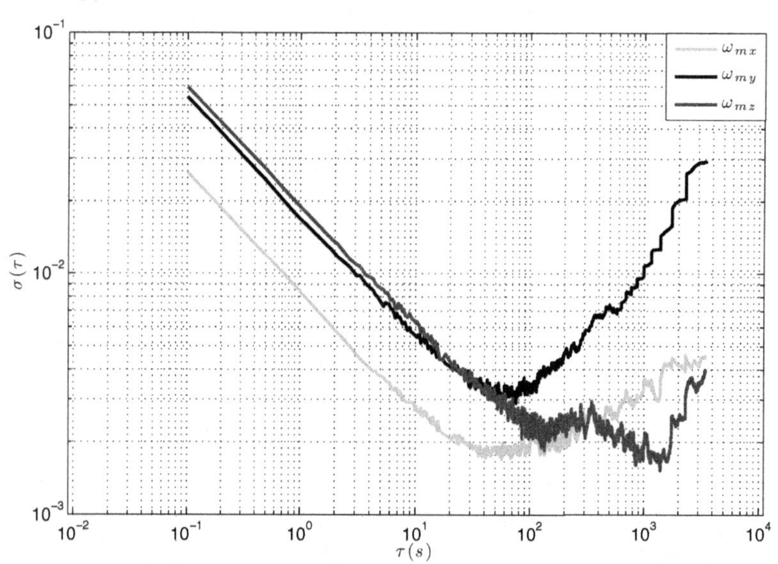

(b) Allan variance associated with the gyrometer measurements

Figure 5.23. *Allan variance associated with each axis of the accelerometer and the gyrometer. For a color version of this figure, see www.iste.co.uk/condomines/kalman.zip*

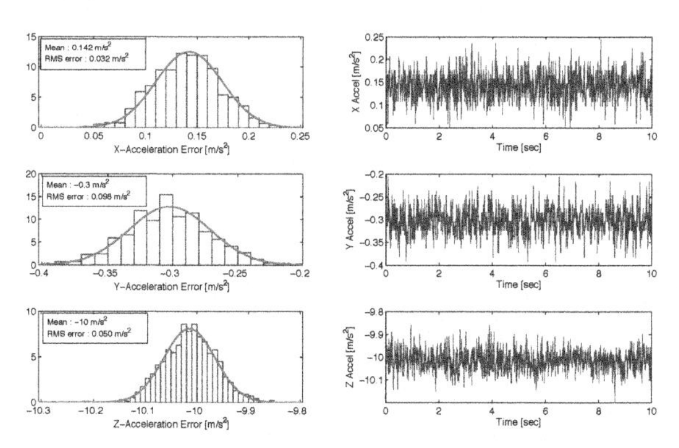

(a) Distribution and standard deviation of the measurement noise in the accelerometer

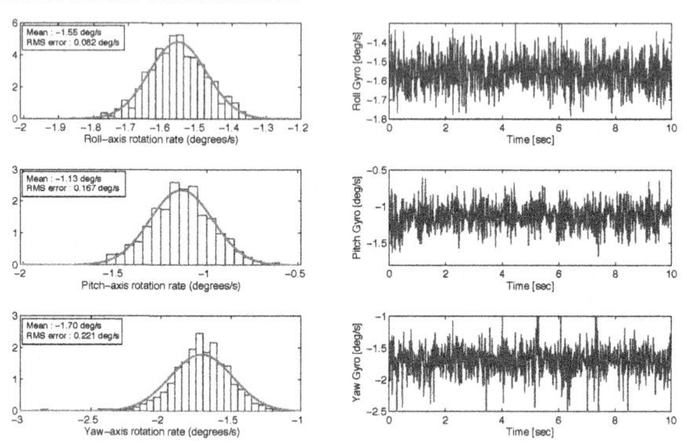

(b) Distribution and standard deviation of the measurement noise in the gyrometer

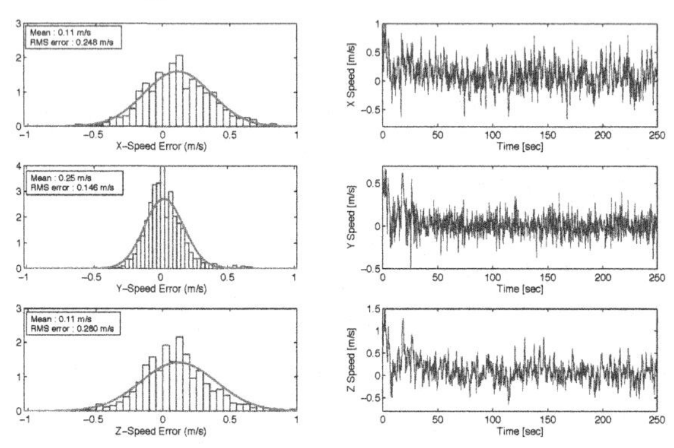

(c) Distribution and standard deviation of the measurement noise in the velocity

Figure 5.24. *Distributions and standard deviations of the measurement noise. For a color version of this figure, see www.iste.co.uk/condomines/kalman.zip*

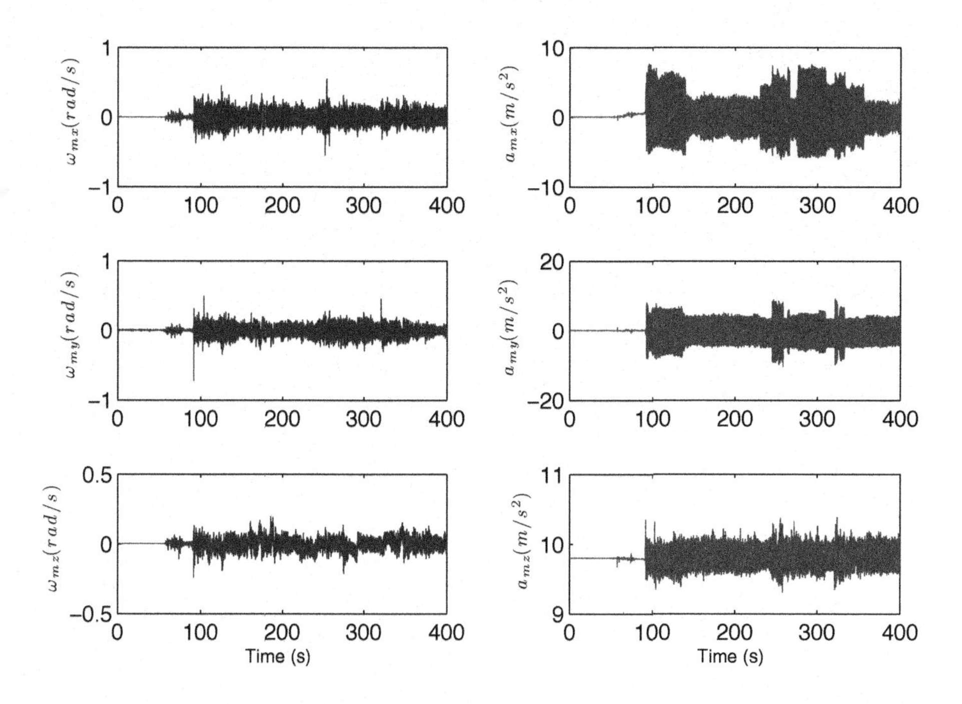

Figure 5.25. *Angular velocity and acceleration measurements*

5.2.3. *Calibration procedure for the embedded sensors of the Apogee chip*

Before we can reconstruct the state from this real data, we need to find calibration values that ensure that each sensor produces consistent output values. This step also accounts for any characteristic properties of the environment, such as the magnetic field at the flight location. The procedure determines a set of deterministic biases and scale factors for each sensor in the IMU. A general model for each axis of the sensor can be written as follows:

$$\begin{pmatrix} \zeta_{m_x} \\ \zeta_{m_y} \\ \zeta_{m_z} \end{pmatrix} = \begin{pmatrix} S_{f_x} & 0 & 0 \\ 0 & S_{f_y} & 0 \\ 0 & 0 & S_{f_z} \end{pmatrix} \cdot \left[\begin{pmatrix} \zeta_{t_x} \\ \zeta_{t_y} \\ \zeta_{t_z} \end{pmatrix} - \begin{pmatrix} b_x \\ b_y \\ b_z \end{pmatrix} \right], \qquad [5.4]$$

where ζ_m is the measurement produced by the sensor and ζ_t is the "true" measurement at time t. The parameters S_f and b, respectively, denote a scale factor and a deterministic bias; the values of these parameters need to be determined for each component of the measurement.

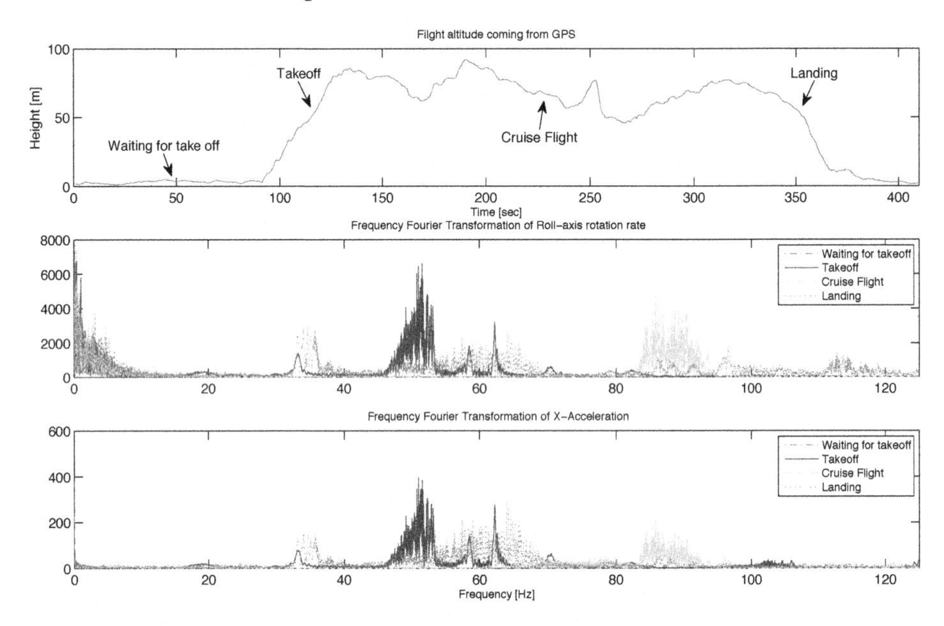

Figure 5.26. *Fourier spectrum of the signals ω_{mx} and a_{mx} in each flight phase. For a color version of this figure, see www.iste.co.uk/condomines/kalman.zip*

5.2.3.1. *Calibration of the accelerometers*

We can calibrate the accelerometers as follows. Using equation 5.4, we can project the acceleration vector A from ground coordinates into aircraft coordinates by performing

$$q^{-1} * A * q = \begin{pmatrix} S_{f_{ax}} & 0 & 0 \\ 0 & S_{f_{ay}} & 0 \\ 0 & 0 & S_{f_{az}} \end{pmatrix} \cdot \left[\begin{pmatrix} \zeta_{t_{ax}} \\ \zeta_{t_{ay}} \\ \zeta_{t_{az}} \end{pmatrix} - \begin{pmatrix} b_{ax} \\ b_{ay} \\ b_{az} \end{pmatrix} \right], \qquad [5.5]$$

where $A = (0\ 0\ -9.81)^T$ is the vector of acceleration due to gravity expressed in ground coordinates. Note that equation 5.5 changes the orientation of the vehicle while preserving the norm of its vectors. Thus, we can calibrate by taking the norm of this equation to obtain a scalar equation that no longer depends on the orientation of the vehicle. This yields the following relation, which involves a total of six calibration variables:

$$9.81^2 = (S_{f_{ax}}(\zeta_{t_{ax}} - b_{ax}))^2 + (S_{f_{ay}}(\zeta_{t_{ay}} - b_{ay}))^2 + (S_{f_{az}}(\zeta_{t_{az}} - b_{az}))^2. \quad [5.6]$$

By recording a large number of measurements with the UAV in various positions, we can calculate the set of six calibration variables with average norm closest to the gravity vector. In Paparazzi, this operation is performed by an optimization algorithm implemented in the programming language, Python.

5.2.3.2. *Calibration of the magnetometers and gyrometers*

As noted in Chapter 3, the Earth's magnetic field is different depending on our location on Earth. Consequently, we need to recalibrate the magnetometer whenever we plan to fly to/in a new location. Unlike for accelerometers, considering the norm at initialization is not enough; we need to know the value of each component. After recording a large number of measurements while varying the orientation of the UAV, we can use the same optimization algorithm used for the accelerometer to graphically project each measurement onto the unit sphere (see Figure 5.27). This allows us to find the scale factor and biases for each component.

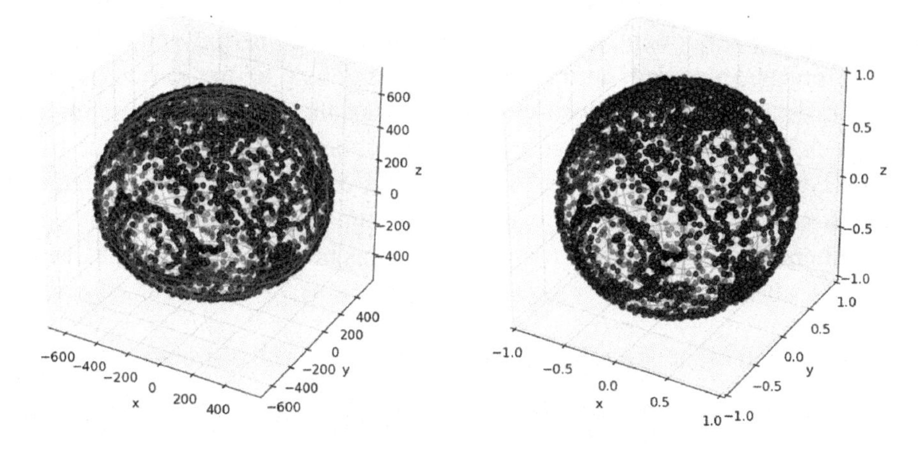

Figure 5.27. *Calibration of the magnetometers. For a color version of this figure, see www.iste.co.uk/condomines/kalman.zip*

Finally, to calibrate the gyrometers, we simply need to compute the deterministic bias in each component $b_{g_{(x,y,z)}}$ as follows:

$$b_{mg_{(x,y,z)}} = \frac{1}{N_s} \sum_{k=1}^{N_s} \zeta_{mg}(k), \qquad [5.7]$$

where N_s is the number of samples and $\zeta_{mg_{(x,y,z)}}(k)$ is the measurement of the component x, y or z at time k.

5.2.4. *Experimental implementation of the π-IUKF algorithm on the AHRS model*

To investigate the global performance of the SR-UKF and π-IUKF algorithms on real data, we chose a flight plan for which the amplitudes of the attitude angles, the position and the velocity would be sufficiently large to evaluate each of the estimation techniques. The flight consists of four phases over a total period of 140 s: (1) waiting for take-off $(0 < t < 50\,\text{s})$; (2) take-off $(72 < t < 80\,\text{s})$; (3) a circuit $(80 < t < 120\,\text{s})$; (4) landing $(120 < t < 140\,\text{s})$. We begin by studying the convergence of the bias terms during the waiting phase. In this phase, the UAV is at rest for 50 s, with its motor turned off, oriented arbitrarily in space. The inertial and magnetic measurements recorded during this period are sent to the various filters described in the previous section to reconstruct the attitude of the aircraft. Figure 5.28 shows the convergence patterns of the estimated bias to the true bias reconstructed from the recorded data. This confirms the results obtained in section 5.1.2, ultimately providing good estimated values for the bias for both filters. There are slightly more perturbations in the transient phase of the SR-UKF filter.

The second flight phase is more dynamic; control of the aircraft was taken over between 54 and 60 s, and the motor was turned on a few seconds before the aircraft was launched at $t = 72$ s. The aircraft was then flown in a closed circuit over the period between 80 and 120 s before finally landing on the ground at $t = 140$ s. The global performance of the attitudes estimated by the SR-UKF and π-IUKF algorithms is plotted in Figure 5.29. The estimates constructed by both algorithms were extremely close throughout the entire dynamic phase. The characteristic features of this phase are clearly visible:

the take-off spike at $t = 72\text{s}$, visible in the roll and pitch axes; the changes in the heading while the aircraft is performing its closed circuit; and finally the landing, which ended with a yaw angle almost identical to take-off, showing that the aircraft returned to its initial position as expected.

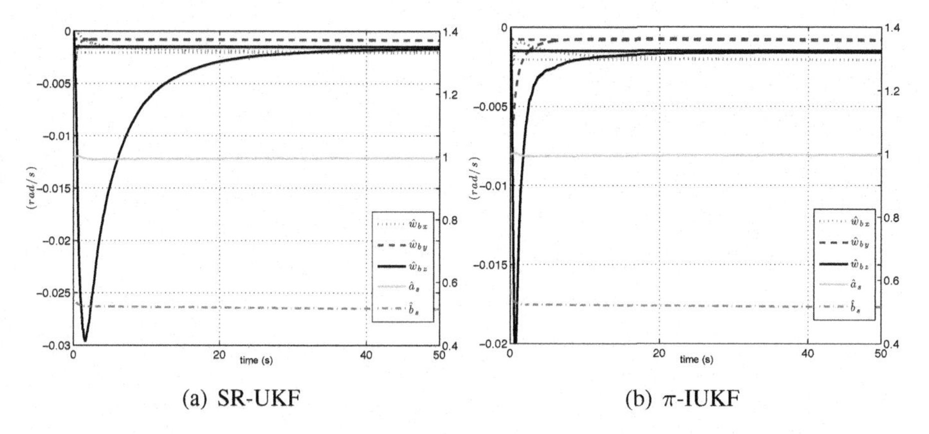

(a) SR-UKF (b) π-IUKF

Figure 5.28. *Convergence of the bias terms and scale factors computed by the SR-UKF and π-IUKF algorithms. For a color version of this figure, see www.iste.co.uk/condomines/kalman.zip*

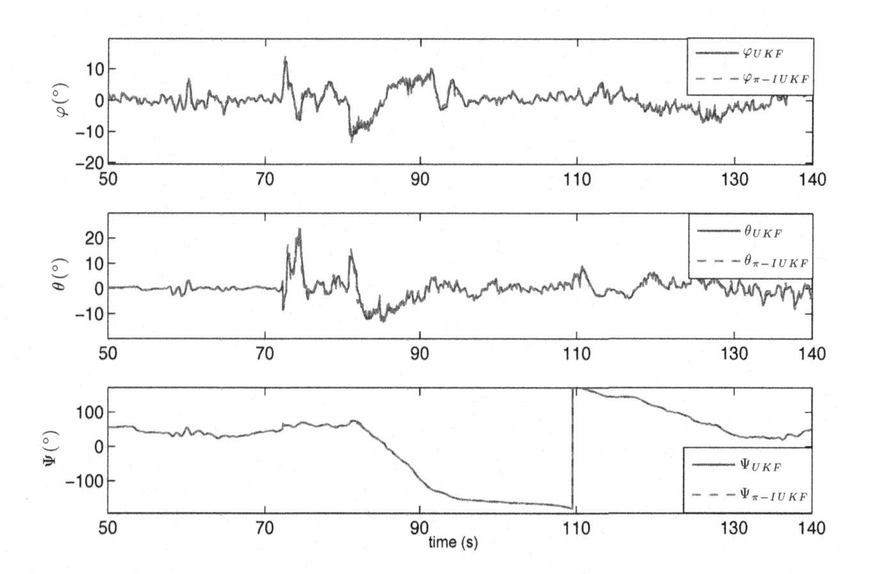

Figure 5.29. *Estimated attitude angles. For a color version of this figure, see www.iste.co.uk/condomines/kalman.zip*

We must now study the convergence of the gains of the π-IUKF and UKF filters to validate the invariance properties of the newly proposed algorithm on real data. Figure 5.30 lists the various matrix coefficients $\mathbf{K}$ constructed by SR-UKF and π-IUKF. The invariance properties of π-IUKF are indeed visible in the results, and so the same conclusions hold as in section 5.1.3. First, the gain terms of both filters are constant during the waiting phase, which is a constant trajectory. The coefficients computed by both algorithms clearly vary during the phase between 55 and 60 s as the motors were turned on; in this phase, the noise takes on a transient character and degrades the invariance properties for a short period of time. In subsequent phases, the coefficients of the UKF algorithm vary over time, whereas the coefficients of π-IUKF remain constant over every nearly constant trajectory defined by $I(\hat{x}, u) = e$. This demonstrates two things: first, both filters perform similarly at reconstructing the orientation of the aircraft in each flight phase; second, the new algorithm converges on a larger domain than UKF. In future projects, the invariance properties of π-IUKF can be exploited to deduce constant gain values from the data. In this way, the characteristics of each state variable can be used to implement fault-monitoring functionality.

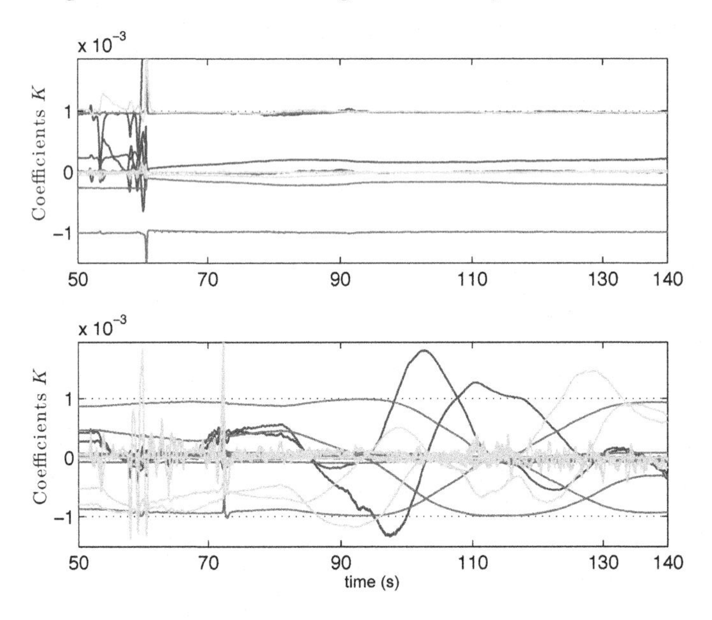

Figure 5.30. *Convergence of the coefficients K computed by the SR-UKF and π-IUKF algorithms. For a color version of this figure, see www.iste.co.uk/condomines/kalman.zip*

5.2.5. *Validation of the INS with real data*

In this section, we present a comparative study of the attitude estimation performance of the UKF and π-IUKF algorithms during navigation. Both approaches are applied to a complete INS model consisting of 15 state variables. First, we study the global performance of both algorithms to validate the quality of the estimates in a real setting. We also present the convergence of the gain terms constructed by both algorithms in order to investigate the invariance properties of π-IUKF for the INS problem based on real data. Next, we take a closer look at the estimation performance of all three proposed algorithms for the velocity and the position by comparing the estimates against reference data recorded by a DGPS antenna. As before for the AHRS, we begin by studying the convergence of the bias terms during the waiting phase, where the UAV is left immobile with its motor turned off for 50 s, oriented arbitrarily in space. Figure 5.31 presents the convergence of the estimated bias to the real bias, reconstructed from the recorded data. Note that the bias in the barometer is discussed later when we consider the dynamic flight phase. The results show that the transient phase is much noisier for the INS model by comparison with the AHRS model, but the convergence is much quicker (around 20 s). The global performance of the estimated attitude angles is plotted on Figure 5.32 for the SR-UKF and π-IUKF filters. The estimates produced by both algorithms are extremely close throughout the entire dynamic phase of the system. However, the results are slightly different from the results presented earlier for the AHRS model. The discrepancy may have been caused by unmodeled imperfections in the system.

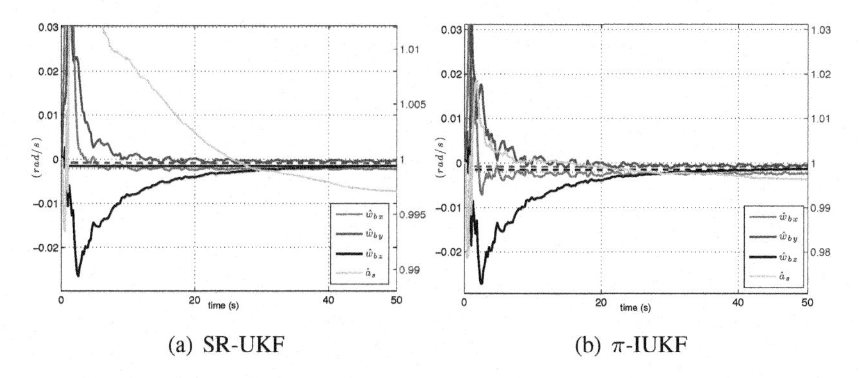

(a) SR-UKF (b) π-IUKF

Figure 5.31. *Convergence of the bias terms of the SR-UKF and π-IUKF algorithms. For a color version of this figure, see www.iste.co.uk/condomines/kalman.zip*

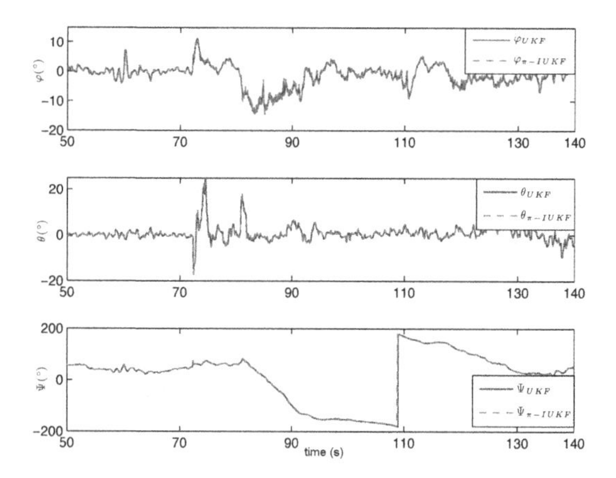

Figure 5.32. *Estimated attitude angles. For a color version of this figure, see www.iste.co.uk/condomines/kalman.zip*

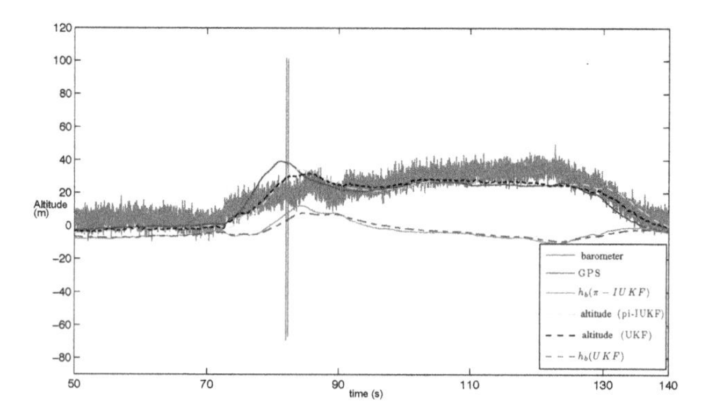

Figure 5.33. *Estimated altitude. For a color version of this figure, see www.iste.co.uk/condomines/kalman.zip*

Figures 5.34 and 5.35 represent the gain terms constructed by SR-UKF and π-IUKF. More precisely, Figure 5.34 lists the coefficients representing the quaternion. The components of the quaternion are no longer all close to constant, unlike section 5.1.5. This may be because the real data includes unsteady noise, contradicting the hypothesis that the noise introduced into the G-invariant system is itself invariant. Figure 5.25 clearly shows variations in the amplitude of the noise during each flight phase, which directly affects the set of gain terms. In conclusion, the gain variations are caused by the fact that the state estimation error is no longer autonomous like the AHRS model, as well as the presence of unsteady noise terms that only appear in experimental settings.

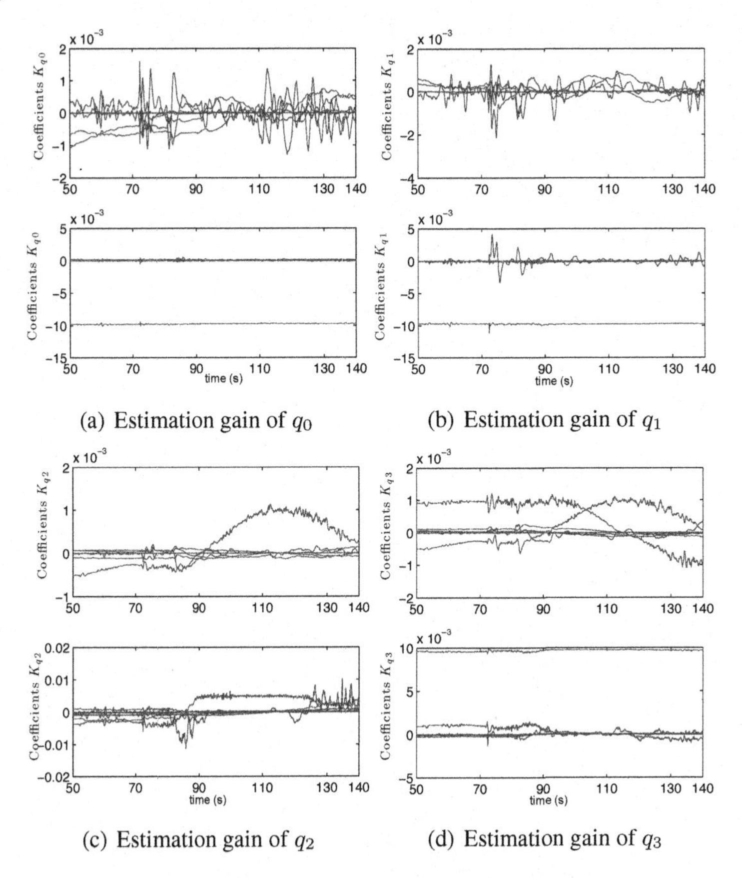

(a) Estimation gain of q_0

(b) Estimation gain of q_1

(c) Estimation gain of q_2

(d) Estimation gain of q_3

Figure 5.34. *Convergence of the estimation gain terms of the quaternion. For a color version of this figure, see www.iste.co.uk/condomines/kalman.zip*

Figure 5.33 shows the altitude measurements from the barometer and the GPS, as well as the result of merging the two measurements. This illustrates a special case only encountered during experiments involving real flights – the barometer measurement is extremely noisy, whereas the GPS measurement gives a particularly good reading of the altitude. Each flight phase defined in the previous section can be distinguished. As the aircraft rises in altitude up to $t = 85$ s, the barometer gives an incorrect reading and the GPS measurement seems to indicate that the altitude increases sharply. Here, the biases of the two algorithms compensated the difference between the barometer and the GPS, since the barometer did not seem to register that the aircraft was gaining altitude. Thus, the SR-UKF and π-IUKF algorithms mitigated any danger to the aircraft caused by incorrect barometer readings, which might have caused the estimates to diverge or led to a loss of control of the aircraft if only the barometer measurement were available. There is an obvious question that needs to be answered: which reading is correct? The barometer or the GPS? To find out, we used a high-accuracy antenna to obtain reference measurements for the position and the velocity in order to study the performance of the two algorithms more closely.

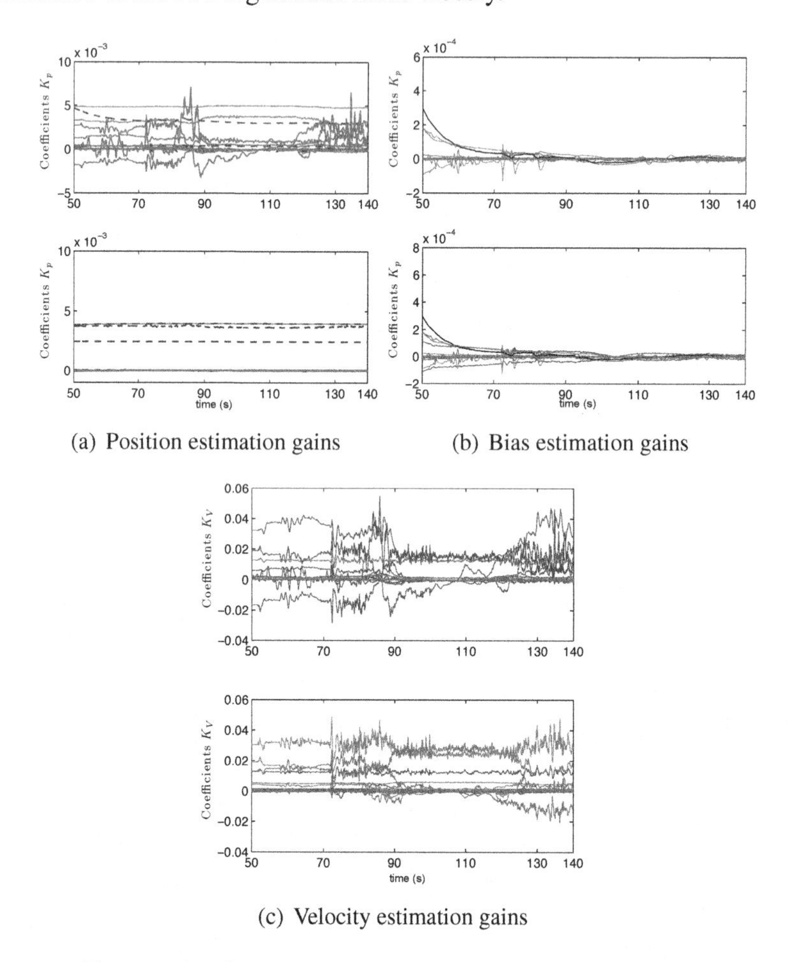

(a) Position estimation gains (b) Bias estimation gains

(c) Velocity estimation gains

Figure 5.35. *Convergence of the position, bias and velocity estimation gains. For a color version of this figure, see www.iste.co.uk/condomines/kalman.zip*

5.2.6. *Analysis of the velocity and position performance of the INS*

This section studies the performance of the velocity and position estimates of the three algorithms proposed in this book more closely by comparing the estimates against reference data from a DGPS. As described at the start of the experimental section of the book, we used a Novatel antenna weighing 150 g to record the reference measurements. Due to its weight, this antenna can only be installed on medium-sized UAVs. We therefore mounted the antenna on a fixed-wing aircraft known as a Twinstar, which is shown in Figure 5.36.

Figure 5.36. *A Twinstar equipped with a Novatel antenna. For a color version of this figure, see www.iste.co.uk/condomines/kalman.zip*

The Apogee autopilot chip allowed us to simultaneously record the GPS data from the Ublox antenna and the data from the Novatel antenna. We performed an additional post-flight processing step with the measurements from the Novatel antenna to obtain highly accurate position data with respect to a local reference system, correlating the DGPS data against a station whose coordinates are very precisely known. Figure 5.37(b) shows the flight altitude, represented in terms of the four flight phases described in section 5.2.2 (waiting for take-off, circuit and landing). This gives another illustration of the estimation performance of each of the three algorithms: the invariant observer, the SR-UKF algorithm and the π-IUKF algorithm. The first thing to note is the mediocre accuracy of the GPS ($\pm 20\,\mathrm{m}$). Each algorithm merges this measurement with the barometer reading. Thus, each algorithm is capable of reconstructing a noise-free estimate, but there nonetheless remains significant error in the altitude.

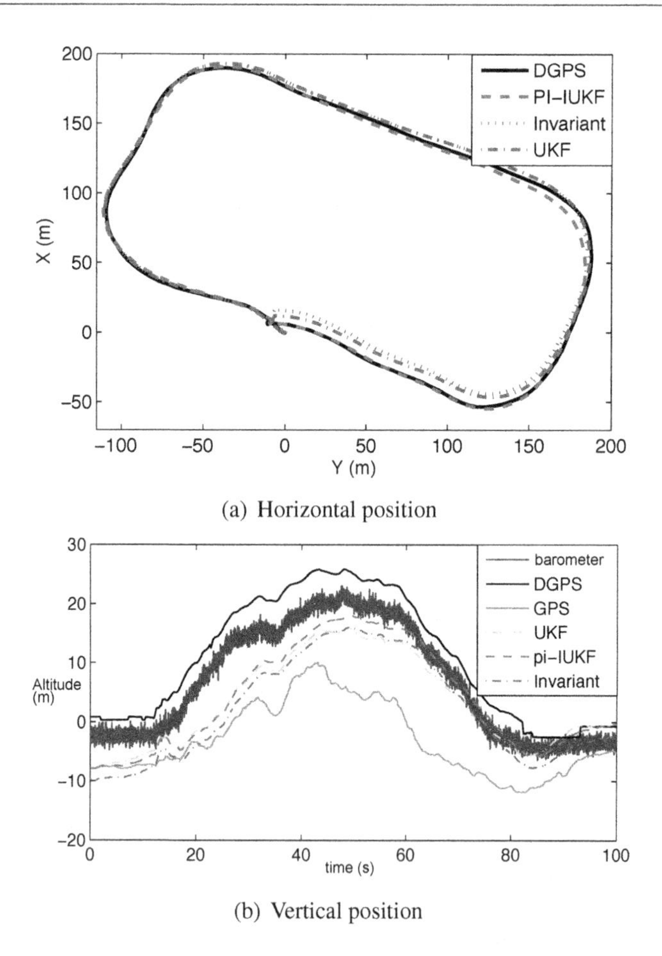

Figure 5.37. *Horizontal and vertical positions of the Twinstar. For a color version of this figure, see www.iste.co.uk/condomines/kalman.zip*

The π-IUKF algorithm gave better altitude estimates than the two other algorithms. Figure 5.37(a) gives a two-dimensional picture of the X and Y positions. The UAV takes off in the direction $X > 0$ and $Y < 0$, performs a circuit and then lands facing $X < 0$ and $Y > 0$. There is some distance between π-IUKF and the two other algorithms in the final phase. The discrepancy is primarily attributable to unmodeled effects, such as the vibrations experienced by the aircraft as it decelerates. Figures 5.37(a) and 5.38 compare the three key algorithms studied throughout this book (the invariant observer, the SR-UKF algorithm and the π-IUKF algorithm), allowing us to conclude that the estimation performance of the π-IUKF algorithm is comparable if not better than the two other algorithms in terms of the position and the velocity. This result is not surprising in light of our earlier conclusions from the simulations. Section 5.3 lists the code of the invariant observer program implemented in the Paparazzi project, written in the programming language C. This code only uses 35% of the CPU capacity of an STM32 microprocessor manufactured by STMicroelectronics.

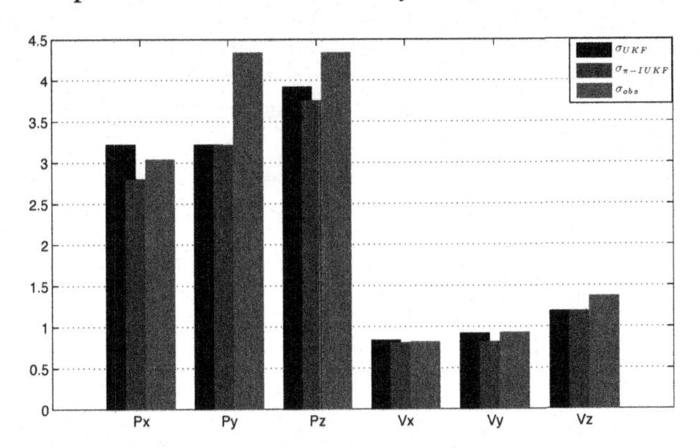

Figure 5.38. *Performance of the three algorithms. For a color version of this figure, see www.iste.co.uk/condomines/kalman.zip*

5.3. Implementation of the invariant observer for the INS model

```
/*
 * Copyright (C) 2012-2013 Jean-Philippe Condomines, Gautier Hattenberger
 *
 * This file is part of paparazzi.
 * paparazzi is free software; you can redistribute it and/or modify
 * it under the terms of the GNU General Public License as published by
 * the Free Software Foundation; either version 2, or (at your option)
 * any later version.
 *
 * paparazzi is distributed in the hope that it will be useful,
 * but WITHOUT ANY WARRANTY; without even the implied warranty of
 * MERCHANTABILITY or FITNESS FOR A PARTICULAR PURPOSE.  See the
 * GNU General Public License for more details.
 *
 * You should have received a copy of the GNU General Public License
 * along with paparazzi; see the file COPYING.  If not, see
 * <http://www.gnu.org/licenses/>.
 *
```

```
 *Compute dynamic mode
 *
 * x_dot = evolution_model + (gain_matrix * error)
 */
static inline void invariant_model(float* o, const float* x,
 const int n, const float* u, const int m __attribute__((unused))) {

  /* dot_q = 0.5 * q * (x_rates - x_bias) + LE * q + (1 - ||q||^2) * q */
  RATES_DIFF(rates_unbiased, c->rates, s->bias);
  /* qd = 0.5 * q * rates_unbiased = -0.5 * rates_unbiased * q */
  float_quat_derivative(&s_dot.quat, &rates_unbiased, &(s->quat));

  float_quat_vmul_right(&tmp_quat, &(s->quat), &ins_impl.corr.LE);
  QUAT_ADD(s_dot.quat, tmp_quat);

  float norm2_r = 1. - FLOAT_QUAT_NORM2(s->quat);
  QUAT_SMUL(tmp_quat, s->quat, norm2_r);
  QUAT_ADD(s_dot.quat, tmp_quat);

  /* dot_V = A + (1/as) * (q * am * q-1) + ME */
  struct FloatQuat q_b2n;
  float_quat_invert(&q_b2n, &(s->quat));
  float_quat_vmult((struct FloatVect3*)&s_dot.speed, &q_b2n, &(c->accel));
  VECT3_SMUL(s_dot.speed, s_dot.speed, 1. / (s->as));
  VECT3_ADD(s_dot.speed, A);
  VECT3_ADD(s_dot.speed, ins_impl.corr.ME);

  /* dot_X = V + NE */
  VECT3_SUM(s_dot.pos, s->speed, ins_impl.corr.NE);

  /* bias_dot = q-1 * (OE) * q */
  float_quat_vmult(&tmp_vect, &(s->quat), &ins_impl.corr.OE);
  RATES_ASSIGN(s_dot.bias, tmp_vect.x, tmp_vect.y, tmp_vect.z);

  /* as_dot = as * RE */
  s_dot.as = (s->as) * (ins_impl.corr.RE);

  /* hb_dot = SE */
  s_dot.hb = ins_impl.corr.SE;

    // set output
  memcpy(o, &s_dot, n*sizeof(float));

/** Compute correction vectors
 * E = ( ? - y )
 * LE, ME, NE, OE : ( gain matrix * error )
 */
```

```
static inline void error_output(struct InsFloatInv * _ins) {

  struct FloatVect3 YBt, I, Ev, Eb, Ex, Itemp, Ebtemp, Evtemp;
  float Eh;
  float temp;

  // test accel sensitivity
  if (fabs(_ins->state.as) < 0.1) {
    // too small, don't do anything to avoid division by 0
    return;
  }

  /* YBt = q * yB * q-1  */
  struct FloatQuat q_b2n;
  float_quat_invert(&q_b2n, &(_ins->state.quat));
  float_quat_vmult(&YBt, &q_b2n, &(_ins->meas.mag));

  float_quat_vmult(&I, &q_b2n, &(_ins->cmd.accel));
  VECT3_SMUL(I, I, 1. / (_ins->state.as));

  /*--------- E = ( ? - y ) ----------*/
  /* Eb = ( B - YBt ) */
  VECT3_DIFF(Eb, B, YBt);

  // pos and speed error only if GPS data are valid
  // or while waiting first GPS data to prevent diverging
  if ((gps.fix == GPS_FIX_3D && ins.status == INS_RUNNING
#if INS_UPDATE_FW_ESTIMATOR
    && state.utm_initialized_f
#else
    && state.ned_initialized_f
#endif
    ) || !ins_gps_fix_once) {
    /* Ev = (V - YV)   */
    VECT3_DIFF(Ev, _ins->state.speed, _ins->meas.speed_gps);
    /* Ex = (X - YX)  */
    VECT3_DIFF(Ex, _ins->state.pos, _ins->meas.pos_gps);
  }
  else {
    FLOAT_VECT3_ZERO(Ev);
    FLOAT_VECT3_ZERO(Ex);
  }
  /* Eh = < X,e3 > - hb - YH */
  Eh = _ins->state.pos.z - _ins->state.hb - _ins->meas.baro_alt;

  /*--------------Gains--------------*/
  /**** LvEv + LbEb = -lvIa x Ev +  lb < B x Eb, Ia > Ia *****/
```

```
VECT3_SMUL(Itemp, I, -_ins->gains.lv/100.);
VECT3_CROSS_PRODUCT(Evtemp, Itemp, Ev);

VECT3_CROSS_PRODUCT(Ebtemp, B, Eb);
temp = VECT3_DOT_PRODUCT(Ebtemp, I);
temp = (_ins->gains.lb/100.) * temp;

VECT3_SMUL(Ebtemp, I, temp);
VECT3_ADD(Evtemp, Ebtemp);
VECT3_COPY(_ins->corr.LE, Evtemp);

/***** MvEv + MhEh = -mv * Ev + (-mh * <Eh,e3>)********/
_ins->corr.ME.x = (-_ins->gains.mv) * Ev.x + 0.;
_ins->corr.ME.y = (-_ins->gains.mv) * Ev.y + 0.;
_ins->corr.ME.z = ((-_ins->gains.mvz) * Ev.z) + ((-_ins->gains.mh)
          * Eh);

/****** NxEx + NhEh = -nx * Ex + (-nh * <Eh, e3>) ********/
_ins->corr.NE.x = (-_ins->gains.nx) * Ex.x + 0.;
_ins->corr.NE.y = (-_ins->gains.nx) * Ex.y + 0.;
_ins->corr.NE.z = ((-_ins->gains.nxz) * Ex.z) + ((-_ins->gains.nh)
          * Eh);

/****** OvEv + ObEb = ovIa x Ev - ob < B x Eb, Ia > Ia ********/
VECT3_SMUL(Itemp, I, _ins->gains.ov/1000.);
VECT3_CROSS_PRODUCT(Evtemp, Itemp, Ev);

VECT3_CROSS_PRODUCT(Ebtemp, B, Eb);
temp = VECT3_DOT_PRODUCT(Ebtemp, I);
temp = (-_ins->gains.ob/1000.) * temp;

VECT3_SMUL(Ebtemp, I, temp);
VECT3_ADD(Evtemp, Ebtemp);
VECT3_COPY(_ins->corr.OE, Evtemp);
/* a scalar */
/****** RvEv + RhEh = rv < Ia, Ev > + (-rhEh) **************/
_ins->corr.RE = ((_ins->gains.rv/100.) * VECT3_DOT_PRODUCT(Ev, I)) +
 ((-_ins->gains.rh/10000.) * Eh);
/****** ShEh ******/
_ins->corr.SE = (_ins->gains.sh) * Eh;
```

Conclusion and Outlook

The objective of this book is to propose new algorithmic solutions to the problem of estimating the state of a mini-UAV during flight, that are compatible with the inherent payload constraints of the Paparazzi system. Recent developments prompted us to focus our research on nonlinear estimation methods based on a pair of suitable models. In particular, we explored two estimation techniques in further detail: Kalman filtering and invariant observers. The "unscented" Kalman filtering technique developed by Julier and Uhlmann proved to be especially valuable numerically. The final months of our research were dedicated to implementing this algorithm into the Paparazzi autopilot. The algorithm uses 60% of the CPU capacity of an STM32 microprocessor manufactured by STMicroelectronics. This is satisfactory in terms of the computational load, which suggests that the other algorithmic solutions developed throughout these pages might also be practically implementable in future.

We were also inspired to study the theoretical concepts associated with invariant observers. Differential geometry, which offers useful insight into the underlying methodology of these estimators, proved especially elusive. We attempted to present the relevant ideas as accessibly as possible in this book. After moving beyond the purely theoretical framework of invariant observers, there is in fact a straightforward method of constructing observers for nonlinear systems with symmetry properties, as demonstrated by our example of a non-holonomic car. However, this method does not provide a systematic strategy to find useful symmetries. In the context of our application, the invariance properties of the kinematic equations can be interpreted physically in terms of the Lie group and the dynamics of the motion of the aircraft. For systems with high levels of intrinsic complexity, interpretations of the dynamics can be more difficult to establish. As an example, this is likely to be the case whenever the flight dynamics are nonlinear.

Over the course of our research, we developed two nonlinear estimation algorithms that combine the theory of invariant observers with the principles of unscented Kalman filtering. We called these algorithms IUKF and π-IUKF. Starting from the same principles and computation steps as the standard UKF algorithm, the usual equations can be adapted to an invariant context, by either:

– redefining the error terms of the standard version of the UKF algorithm – this gives the IUKF algorithm; or

– introducing a compatibility condition – this gives π-IUKF.

These two estimators provide extremely promising algorithmic solutions to the some of the problems encountered by inertial navigation. The invariant state estimation error makes it much easier to analyze the convergence of the estimating filter. After specifying the target dynamics, each method can be used to systematically compute numerical values for the correction gain terms to equip the filter with these dynamics while ensuring that the domain of convergence of the invariant state estimation error remains large. The various results from simulated data and real measurements presented earlier showed that the invariant state error only depends on the estimated trajectory via the invariants of the problem, which effectively act like a nonlinear function.

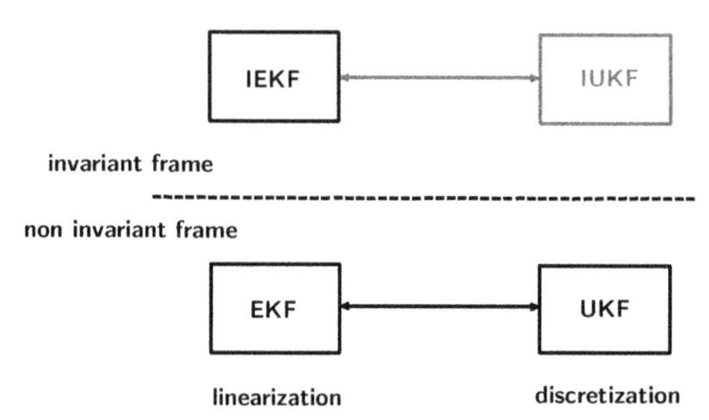

Figure C.1. *Reformulation of the UKF algorithm for an invariant setting*

Thus, whenever the estimated trajectory of the system is close to constant, the correction gains and the state covariance errors converge numerically to constant values. Our results showed that this was in particular true when the algorithms were applied to an AHRS model, which satisfies the property that the invariant state error is autonomous. By contrast, for the INS, we found that it was only possible to obtain constant gain terms on some constant trajectories. For example, Bonnabel *et al.* [BON 08] showed that the trajectories stabilize when the mini-UAV engages in uniformly accelerated helicoidal motion along an axis of rotation and accelerated translational motion in a straight line. For both navigation models, it is important to note that the invariance properties are slightly modified when unsteady noise is introduced into the system, as can be seen in our results on real data. One possible solution to this obstacle might be to mechanically attenuate the vibrations by mounting the autopilot card on a suspension system relative to the body of the aircraft.

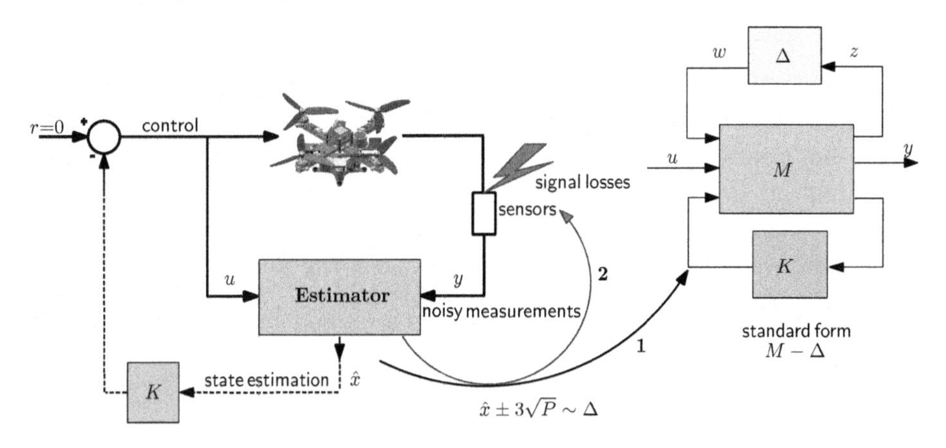

Figure C.2. *Improving the control of nonlinear systems*

Although the initial results of the theoretical work presented here have been promising, there is still plenty of potential for improvement that remains to be explored, at the level of both methods and applications. From the perspective of *methodology*, the formulation of SR-IUKF for a complete inertial navigation model needs to be evaluated in order to determine whether it could potentially be more efficient than classical UKF, as we did for the π-IUKF algorithm.

Additionally, we still need to compare the performance of SR-IUKF and π-IUKF to find out which is the most suitable for implementation on a mini-UAV. Our research also needs to be extended by using the newly proposed concepts to improve the design of the command loop and to implement fault detection functionality. Given that the state covariance errors constructed by the estimating filters are constant for the AHRS, we have a choice between two possible approaches:

– we could use this information to improve the control of nonlinear systems (Youla parametrization, LQG, H_∞, etc.). Systems based on a Linear Quadratic Gaussian (LQG) controller would be particularly suitable, since the state variable noise terms are normally distributed. In the case of a robust controller, we could also attempt to take into account constant uncertainties in the estimated state variables using an $M - \Delta$ standard form (see Figure C.2);

– we could design a closed-loop system that is robust against some of the issues encountered during mini-UAV flight. For example, there are often significant errors in the accelerometric and gyrometric measurements. These measurements appear in the observation equations of the AHRS and elsewhere. In this scenario, we could actively monitor the quality of the estimates by using the covariance of the state errors to detect any faults that might arise (see Figure C.3).

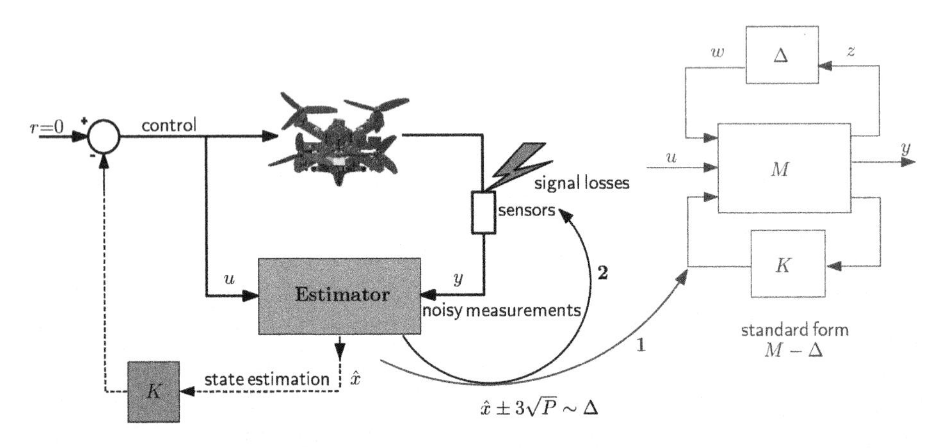

Figure C.3. *Control loop that is robust against potential faults*

From the perspective of *applications*, we have already implemented the invariant observer algorithm into Paparazzi. We applied the techniques presented in Chapter 5 to configure an estimator that produces satisfactory flight behavior from the mini-UAV. The computational load was measured at 35% of the CPU capacity of an STM32 microprocessor manufactured by STMicroelectronics. Although this algorithm is computationally lighter than UKF, its observer does not yet achieve the desired improvements, since it does not estimate the covariance of the state errors. Furthermore, given the tendency of miniaturization and embedded processing power to advance in leaps and bounds, we expect that within two years the computational load of the UKF algorithm will be comparable to that of the invariant observer today. Therefore, our future research will focus on implementing the π-IUKF and IUKF algorithms by reworking the existing code of UKF.

Appendix

Differential Geometry and Group Theory

This appendix gives an intuitive review of a few ideas from differential geometry and group theory needed to properly understand the underlying methodology of the construction of invariant observers. Specifically, we will introduce the notions of manifold, submanifold, vector field and Lie group.

A.1. Manifold and diffeomorphism

Intuitively, the concept of a manifold can be understood as a generalization of the metric and differential properties of "standard" surfaces in Euclidean space to "curved" spaces. Instead of thinking of these new spaces as embedded in some larger higher dimensional space, we view them as reference coordinate spaces. The sphere, the torus and the hyperbolic paraboloid are all examples of two-dimensional manifolds, since the neighborhood of every point on their surfaces can be characterized by a system of two local coordinates (for instance in the case of the sphere, we can specify each point by its longitude and latitude). We can further define a topology or topological space on any manifold, which allows us to classify them as mathematical objects. For example, from a global perspective, the sphere and the torus are not the same. Even though the torus and the sphere have the same number of dimensions, we cannot find a regular mapping between their points as mathematical objects. This fundamental non-equivalence, which is topological in nature, has implications for the global properties of the tangent vector field. Thus, on a torus, we can construct a regular field of non-vanishing tangent vectors, whereas it is impossible to do so on a sphere; this is known as the "hairy ball" or "hedgehog" theorem [BER 72].

For us, a manifold is a mathematical space on which we can study and describe the geometric properties that characterize the dynamics of a physical system (rotation, translation, dilation, etc.) in order to find invariance relations, which will allow us to construct our so-called "invariant" observers. To allow us to apply classical mathematical properties to the results derived in this space, we shall assume that the manifolds are differentiable, or smooth.

More formally, a manifold $\mathcal{M}$ is a collection of open subsets (a topological subspace) $U_\alpha \subset \mathcal{X}$ of the Euclidean space on $\mathbb{R}^m$; thus, the simplest example of a manifold is just a single open subset $\mathcal{X} \subset \mathbb{R}^m$ of the Euclidean space. We can define a chart (i.e. a system of local coordinates) denoted as $\Sigma_\alpha : U_\alpha \mapsto V_\alpha \subset \mathbb{R}^n$ whose coordinates $x^\alpha = (x_1^\alpha, \ldots, x_n^\alpha)$ with respect to the standard basis on V_α determine coordinates for any given point of $\mathcal{X}$. The transition map $\Sigma_{\beta\alpha} = \Sigma_\beta \circ \Sigma_\alpha^{-1} : V_\alpha \mapsto V_\beta$, which is assumed to be continuous, specifies the change of local coordinates between two charts β and α on $\mathcal{X}$. To illustrate this, consider the following example, widely used in various fields of research such as cartography: stereographic projection.

A.1.1. *Illustrative example*

Consider an n-dimensional sphere as a compact topological subspace $\mathbb{S}_n$ of $\mathbb{R}^n$ (see Figure A.1); in other words, an $(n-1)$-dimensional analytic manifold. Choose two points $N = (1, 0, \ldots, 0)$ and $S = (-1, 0, \ldots, 0)$ to be the north and south poles, and write $p_N : \mathbb{S}_n - \{N\} \mapsto \mathbb{R}^n$ and $p_S : \mathbb{S}_n - \{S\} \mapsto \mathbb{R}^n$ for the mappings that send each point x on the sphere distinct from the corresponding pole to the point where the hyperplane with equation $x_0 = 0$ (identified with $\mathbb{R}^n$) intersects with the line passing through that pole and x. These mappings are called stereographic projection from the north and south poles, respectively. Geometrically, they are clearly continuous, bijective and continuously invertible (these properties are summarized by the term "diffeomorphic"). Note that one of the transition maps $p_S \circ p_N^{-1}$ is given by $x \mapsto x/||x||^2$ (since $||p_N(x)|| = tan(\alpha)$ and $||p_S(x)|| = tan(\pi/2 - \alpha)$, so $||p_N(x)|| = 1/||p_S(x)||$).

The manifolds used to construct and study invariant observers are said to be algebraic, which means that they satisfy a polynomial equation of type

$f(x, y, z, \cdots) = 0$. In this context, the notion of local diffeomorphism (a differentiable and continuously invertible bijection from one topological space to another) will play an essential role in constructing descriptions of manifolds that are independent of the choice of coordinates.

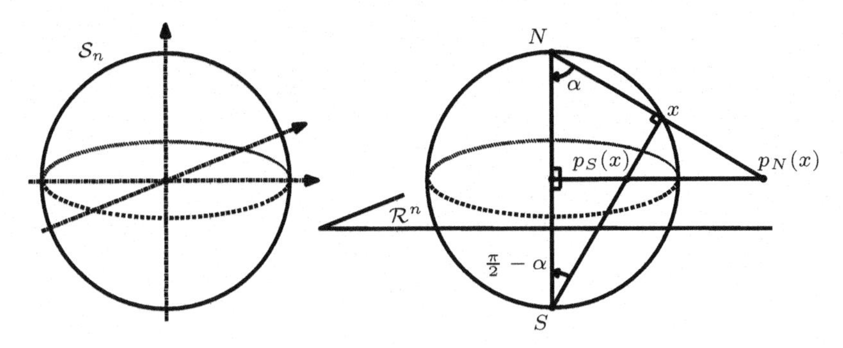

Figure A.1. *Compact topological subspace $\mathbb{S}^n$ of a Euclidean space $\mathbb{R}^n$*

As a result, the methodological results established in this book are only valid when considering a local solution of the topological space that describes the dynamics of the system. To obtain this solution, consider an arbitrary function φ such that $\varphi : \mathbb{R}^n \longmapsto \mathbb{R}^k, n > k$. The level set $S = \{\varphi(x) = c\}$ is said to be regular if it is non-empty and the Jacobian matrix $D\varphi = \partial\varphi^i/\partial x^j$ has rank at most k on S. Thus, each regular level set S forms an analytic submanifold $\mathbb{R}^n$ of dimension $m = n - k$. The implicit function theorem gives a local solution around $(x_1, \ldots, x_k)$ such that $x_i = \varphi_i(x_{k+1}, \cdots, x_n)$, $i = 1, \ldots, k$, giving a local parametrization of the set of level sets as a system of $n - k$ local coordinates. Consider the following example to illustrate this theorem.

A.1.2. *Illustrative example*

Consider the equation of a circle represented by the function $\varphi : \mathbb{R}^2 \mapsto \mathbb{R}$, defined by $\varphi(x_1, x_2) = x_1^2 + x_2^2 - R^2$, where R is a positive real number. This is a one-dimensional analytic submanifold (or embedded manifold). Locally, the circle resembles a line; it is therefore one-dimensional, i.e. one single coordinate suffices to describe small arcs of the circle. Now, consider the top region of the circle ($x_2 > 0$, in yellow on Figure A.2). On this region, we can specify any given point by the coordinate x_1.

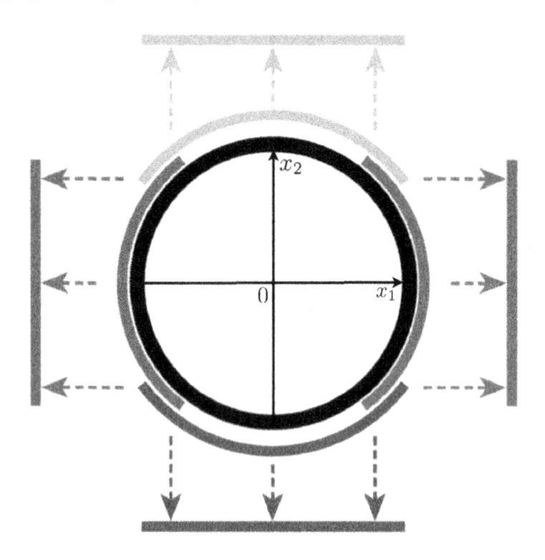

Figure A.2. *Illustration of four charts of the circle S^1. For a color version of this figure, see www.iste.co.uk/condomines/kalman.zip*

This is a chart represented by the local homeomorphism (non-differentiable diffeomorphism) χ_{top}, which sends the yellow region of the circle to the open interval $]-R, R[$ by representing each point on the circle by its first coordinate $\chi_{top}(x_1, x_2) = x_1$. Similarly, there is one chart for each of the bottom, left and right regions of the circle. The trivial solution of the equation $\varphi = 0$ is $x_1 = \pm\sqrt{R^2 - x_2^2}$ for $|x_2| \leq R$, which therefore defines the change of local coordinates for each transition map. The implicit function theorem guarantees the existence of a local solution around any point $(x_{1,0}, x_{2,0})$ such that $x_{1,0}^2 + x_{2,0}^2 = R^2$ (e.g. $x_{1,0} = R, x_{2,0} = 0$), since the tangent linear mapping of φ at this point is $D\varphi(x_{1,0}, x_{2,0}) = (2x_{1,0}, 2x_{2,0}) \neq (0, 0)$, which therefore has rank 1. We will use this theorem again at the end of the appendix to find a local solution for Élie Cartan's method of moving frames.

A.2. Topology and state space

Manifolds, and in particular state manifolds, give a global view of the behavior of a physical system over large intervals of time. This allows us to structure the configuration space of the system by giving a description of the set of positions that this system is likely to take *a priori*. Moreover, this approach allows us to study interesting sets with both algebraic structure and manifold structure, said to be compatible with one another (studying rotations in a three-dimensional space leads to a two-dimensional manifold and a group). Finally, this geometrical generalization allows us to study and develop certain methods in a manner that is independent of the choice of coordinates in the configuration space. As we noted earlier, a manifold can be viewed as a collection of globally coherent open subsets of $\mathbb{R}^m$ that correspond, via their local coordinates, to small pieces (neighborhoods) of the manifold. The concept of smooth manifold was inspired by the study of curves (one-dimensional manifolds) and surfaces (two-dimensional manifolds). Figure A.3 presents a few canonical examples of one- and two-dimensional manifolds:

– the circle S^1 is the canonical example of a closed curve (periodic orbit), and is therefore encountered very frequently when studying periodic behavior. Here, we give its representation in the plane $\mathbb{R}^2$ in the standard coordinates x and y;

– the cylinder $S^1 \times \mathbb{R}$ is the manifold of natural states of the planar pendulum: each of the points (θ, ω) on the cylinder $S^1 \times \mathbb{R}$ corresponds to a velocity vector tangent to the cylinder (see Figure A.4) at that point. The pendulum is characterized by the angle θ from the downward vertical axis defined up to multiples of 2π, i.e. which takes values in $S^1 = \mathbb{R}/2\pi(Z)$, and the angular velocity $\omega = \dot{\theta}$, which ranges from $-\infty$ to $+\infty$. The dynamics of the pendulum are therefore determined by the tangent vector field of the cylinder;

– the torus $T^2 = S^1 \times S^1$ is the state manifold naturally encountered when studying a double pendulum in the plane formed by two decoupled linear oscillators with frequencies ω_1 and ω_2. Indeed, suppose that the phase space is not the usual plane but instead the torus T^2. Then, each open subset of T^2 can be identified with a small region of the plane. Intuitively, it is clear that the asymptotic regime of the system sweeps the entire phase space over time along helix-shaped paths on this torus (see Figure A.5).

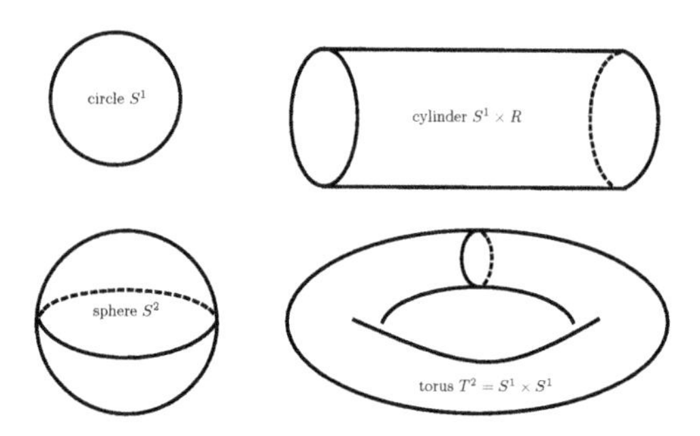

Figure A.3. *Most common examples of state spaces that are straightforward to construct from either the real line $\mathbb{R}$ or the real plane $\mathbb{R}^2$*

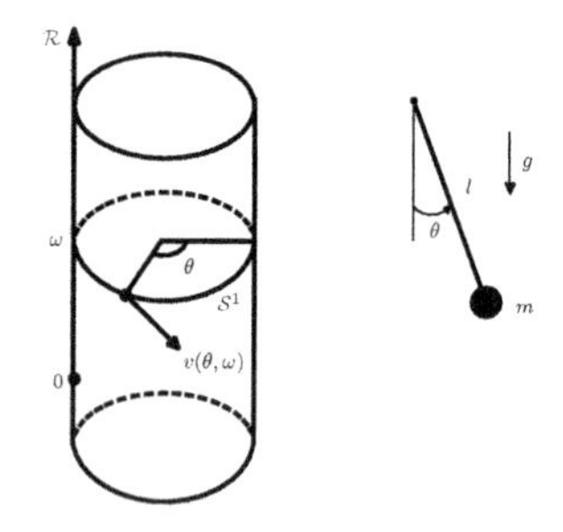

Figure A.4. *The state space of the pendulum is the cylinder parametrized by $(\theta, \dot{\theta} = \omega) \in \mathcal{S}^1 \times \mathbb{R}$. The velocity vector $v(\theta, \omega)$ is tangent to the cylinder*

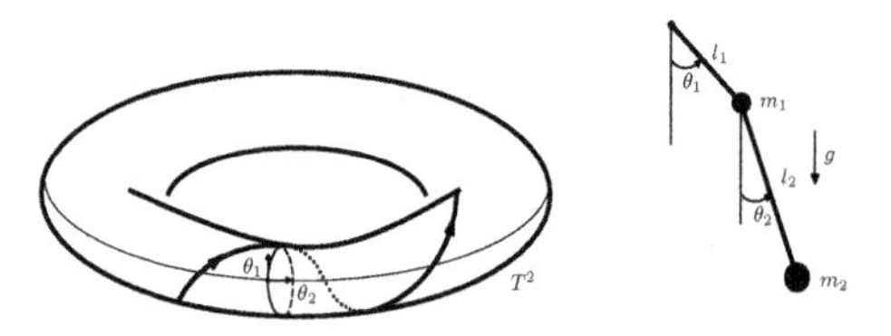

Figure A.5. *The parameters $\dot{\theta}_1 = \omega_1, \theta \in \mathcal{S}^1$ and $\dot{\theta}_2 = \omega_2, \theta_2 \in \mathcal{S}^1$ follow helix-shaped paths on a torus parametrized by the two latitude and longitude angles θ_1 and θ_2*

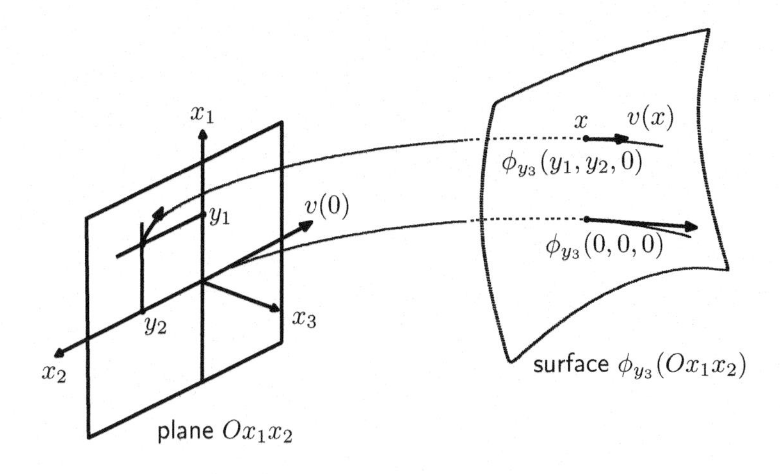

Figure A.6. *Straightening-out theorem in* $\mathbb{R}^3$. *The change in coordinates* $x \to y$ *defined by* $x = (x_1, x_2, x_3) = \phi_{y_3}(y_1, y_2, 0)$ *transforms the field* v *into a constant field*

A.3. Tangent vectors of manifolds and vector fields

Readers may be surprised to discover there is a deep connection between groups and vector fields. This section will need to take a detour via group theory to establish the concept of tangent vector field of a manifold.

The tangent vector to a manifold $\mathcal{M}$ at the point $x \in \mathcal{M}$ is defined geometrically as the tangent vector of any (smooth) curve that passes through x. In local coordinates, the tangent vector $v|_x$ of the curve parametrized by $x = \phi(t)$ is given by the tangent linear mapping $v|_x = D\phi(t)$. Thus, the collection of all tangent vectors forms a tangent space of $\mathcal{M}$ at x, and each tangent space, denoted $TM|_x$, is a vector space with the same number of dimensions as $\mathcal{M}$. In particular, the tangent space of an open set $\mathcal{X}$ at the point $x \in \mathcal{X}$ is defined as the vector space $TX|_x = \ker D\phi(x)$. From the differential point of view, writing $\xi(x) = (\xi^1(x), \cdots, \xi^m(x))$ for the components of ξ, the vector field can be written in local coordinates as:

$$v = \sum_{i=1}^{m} \xi^i(x) \frac{\partial}{\partial x^i}.$$

The curve parametrized by $\phi : \mathbb{R} \longmapsto \mathcal{M}$ is called an integral curve of the vector field v if the tangent vectors $v|_x$ of this curve coincide with the vector field v at every point. This condition only holds if the parametrization $x = \phi(t)$ satisfies a first-order differential equation $dx^i/dt = \xi^i(x)$ for each $i = 1, \ldots, m$. The standard uniqueness and existence theorems for first-order differential systems imply that, given any point $x \in M$, there is a unique maximal integral curve passing through x. We write $\phi(t) = exp(t\xi)x$ for the maximal integral curve passing through $x = exp(0\xi)x$ at $t = 0$. This already gives us our first example of a transformation group generated by a vector field: flows. In the following, we will use the term "infinitesimal generator" instead of vector field for a flow. A flow describes the motion of a point along a curve, i.e. motion with one single degree of freedom.

Finally, anywhere that is not a singularity (a point with value zero), we can locally classify each vector field up to diffeomorphism (e.g. by performing a local change of coordinates). Thus, given any smooth or analytic vector field v on an open set $\mathcal{U}$ and some point $x_0 \in \mathcal{U}$ that is not a singularity, i.e. $v(x_0) \neq 0$, we can "straighten v out" locally. This is known as the straightening-out theorem. Formally, there exists a local diffeomorphism in the neighborhood of $y = (y_1, \cdots, y_n) = f(x)$ that transforms $dx^i/dt = \xi^i(x)$, $i = 1, \cdots, m$, into the following normal form: $(dy_1/dt = 0, dy_{n-1}/dt = 0, \cdots, dy_n/dt = 1)$. To visualize this, simply imagine a hyperplane distinct from $v(0)$ that passes through 0 and whose direction is defined by the vectors $(e_1, \cdots, e_{n-1})$ along each of the first $n-1$ coordinates of x (see Figure A.6). The change in coordinates $x \rightarrow y$ defined by $x = (x_1, x_2, x_3) = \phi_{y_3}(y_1, y_2, 0)$ transforms the field v into a constant field.

A.4. Lie groups

We are interested in one of the most common types of topological space: symmetry groups, more specifically Lie groups, acting on finite-dimensional manifolds. The remarkable field of group theory was founded in the seminal work by the mathematician Sophus Lie in the 20th Century. While traveling in France, Sophus Lie discovered Galois' research on a method that could solve algebraic equations by fixing the set of their roots. Lie was struck by a revelation a few years later: using a process similar to Galois theory, he realized that continuous symmetry groups could help to study differential equations. Since then, a vast theory has been built upon Lie's fundamental insight. This theory, and in particular the notion of infinitesimal generators of continuous groups, has had deep and far-reaching ramifications for every field of mathematics and modern physics. Lie groups are foundational mathematical objects and one of the key components of our approach, so we shall need a basic understanding of what they mean.

A.4.1. *Notion of a group*

Let us begin by recalling that a group is a set G equipped with a binary multiplication operation $g \cdot h$ defined for every pair of elements $g, h \in G$ in the group. Any group must also have a (unique) identity element, denoted e, and every element g in the group must have an inverse g^{-1} such that $g \cdot g^{-1} = g^{-1} \cdot g = e$. The simplest example of a group is the set of real numbers $\mathbb{R}$ under addition. The identity element is 0, and the inverse of any given element x is $-x$. The set of non-zero real numbers $\mathbb{R}^* = \mathbb{R} - \{0\}$ and the set of positive real numbers $\mathbb{R}^+$ are also groups under the separate operation of multiplication, in which case the identity element is the number 1.

Next, let us move on to Lie groups, which are a special case of topological group. A topological group is a group equipped with a topology with respect to which the operations of multiplication and inversion are continuous mappings. Lie groups additionally have an analytic manifold structure with respect to which the group operations $(g, h) \mapsto g \cdot h$ and $g \mapsto g^{-1}$ are analytic charts. Examples of Lie groups include the rotational group $SO(2)$, the group of homotheties and the Euclidean group $SE(2, \mathbb{R})$. An r-dimensional Lie group is typically understood to mean a Lie group with r parameters (or local coordinates $g = (g_1, \ldots, g_r)$). We have already encountered an example of a one-parameter group of diffeomorphisms: the flow of a vector field v on an open set $\mathcal{U} \subseteq \mathbb{R}^m$. As noted, this is indeed a local group with a single parameter $t \cdot x = exp(t\xi)x$.

A.4.2. *Transformations and group actions*

One key area of the theory of invariants is the study of groups acting on manifolds or analytic open sets. The additional structural assumptions provide powerful new analysis tools that would not be available if we considered more general group actions. These group actions allow us to specify an algebra on which concepts relating to both vector fields and differential forms can be applied. In some applications, Lie algebras, or "infinitesimal" Lie groups, are used to replace complicated nonlinear group transformations by a simple infinitesimal equivalent. One extremely interesting example about finding symmetry groups in the heat equations is discussed in [OLV 08]. In this example, the infinitesimal invariance conditions form a system of linear partial differential equations, which allows well-known analytic methods to be applied directly in order to reconstruct the required invariants. The key step in this process is to identify the infinitesimal generators of the group action or transformation as differential operators that act like functions defined on the space.

First, let us present the types of manipulations that can be applied to groups, i.e. group transformations. In general, a transformation group is an analytic mapping from an open set $\mathcal{W}$ to an open set $\mathcal{U}$ such that:

$$\varphi : \begin{cases} (G, \mathcal{W}) & \longrightarrow \mathcal{U} \\ (g, x) & \longmapsto \varphi(g, x) =: g \cdot x \end{cases}$$

We shall consider a "local" group of actions. In other words, given $x \in \mathcal{X}$, the transformation $g \cdot x$ is only defined for elements g in the group that are sufficiently close to the identity element e. Thus, in a local transformation group, the mapping φ is defined on an open set $\mathcal{W}$ such that $\{e\} \times \mathcal{U} \subseteq \mathcal{W} \subseteq G \times \mathcal{U}$ and must satisfy the following properties:

– composition: if (h, x), $(g, h \cdot x)$, and $(g \cdot h, x)$ are in $\mathcal{W}$, then $g \cdot (h \cdot x) = (g \cdot h) \cdot x$;

– identity element: for every x in $\mathcal{U}$, $e \cdot x = x$;

– inverse: the inverse element g^{-1} determines the inverse of the transformation defined by the element g.

For all $g \in G$, the left-multiplication mapping $L_g : G \longmapsto G$ and right-multiplication mapping $R_g : G \longmapsto G$ are defined as follows: $\forall h$, $L_g(h) = g \cdot h$ and $R_g(h) = h \cdot g$. Manipulating transformation groups may seem extremely complex at first. Consider a simple but relevant example: the (real) affine group A(1), which is defined as the transformation group $x \to ax + b$ on the real line $x \in \mathbb{R}$. The affine group is parametrized by the pair (a, b), where a and b are real numbers. The group multiplication operation is defined by $(a, b) \cdot (c, d) = (ac, ad + b)$, and the identity element is $e = (1, 0)$, i.e. $(a, b) \cdot (c, d)|_{c=1,d=0} = (a, b)$. The left-multiplication and right-multiplication charts are therefore given by $R_{(a,b)}(c, d) = (c, d) \cdot (a, b) = (ac, bc + d)$, $L_{(a,b)}(c, d) = (a, b) \cdot (c, d) = (ac, ad + b)$. It can be checked that the right action R may be canonically associated with the left action L by defining $L_{(a,b)}(c, d) = R_{(a,b)^{-1}}(c, d)$. Another important example is the Euclidean group SE(2) acting on itself (from the right) by composition of translations ($x \to x + b$) and rotations (rotation group). This type of action is also known as a group automorphism. The Euclidean group is characterized by the property that its elements are norm-preserving transformations, similar to other isometry groups. For all $(x_0, y_0, \theta_0) \in G$, the mapping $\varphi(x_0, y_0, \theta_0)$, which is a right-automorphism of G, has the following multiplication operation:

$$R_{(a_0,b_0,\theta_0)}(a, b, \theta) = \varphi_{(a_0,b_0,\theta_0)}(a, b, \theta) = \begin{pmatrix} a \\ b \\ \theta \end{pmatrix} \cdot \begin{pmatrix} a_0 \\ b_0 \\ \theta_0 \end{pmatrix}$$
$$= \begin{pmatrix} acos\theta_0 + bsin\theta_0 + a_0 \\ -asin\theta_0 + bcos\theta_0 + b_0 \\ \theta + \theta_0 \end{pmatrix}.$$

This should not be confused with the Euclidean group E(2), which acts on the affine plane $\mathbb{R}^2$ by composition of translations and rotations around the origin. It can also be presented as the semidirect product of the group of rotations $U(1)$ and the group of translations on the plane $\mathbb{R}^2$. For all $(x, y, \theta) \in G$, the mapping $\varphi(x, y, \theta)$ specifying the action of G acts on the affine plane with the following multiplication operation:

$$\varphi_{(a,b,\theta)}(x, y) = \begin{pmatrix} a \\ b \\ \theta \end{pmatrix} \cdot \begin{pmatrix} x \\ y \end{pmatrix} = \begin{pmatrix} xcos\theta - ysin\theta + a \\ xsin\theta + ycos\theta + b \\ \theta + \theta_0 \end{pmatrix}.$$

Finally, note that we can classify Lie groups by their algebraic properties. A list of Lie groups and their algebras is given in Table A.1.

Lie group	Description	Remarks	Lie algebra	Description	dim/$\mathbb{R}$
$\mathbb{R}^n$	Euclidean space under addition	Abelian, non-compact	$\mathbb{R}^n$	Lie bracket is trivial	n
$\mathbb{S}^1$	Unit circle of complex numbers with modulus 1	Space homeomorphic to a circle in the Euclidean plane	$\mathbb{R}$	Lie bracket is trivial	1
$\mathbb{S}^3$	Sphere represented by the quaternions with modulus 1	Simply connected, compact, isomorphic to $SO(3)$	$\mathbb{R}^3$	Quaternions with real part zero; Lie bracket is the vector product	3
$\mathbb{H}^\times$	Non-zero quaternions under multiplication	Simply connected, non-compact	$\mathbb{H}$	Quaternions; Lie bracket is the commutator	4
$GL(n, \mathbb{R})$	General linear group of real invertible $n \times n$ matrices	Non-compact	$M(n, \mathbb{R})$	$n \times n$ matrices; Lie bracket is the commutator	n^2
$SL(n, \mathbb{R})$	Special linear group of real matrices with determinant one	Non-compact if $n > 1$	$\mathfrak{sl}(n, \mathbb{R})$	Square matrices with trace zero; Lie bracket is the commutator	$n^2 - 1$
$O(n, \mathbb{R})$	Orthogonal group of real orthogonal matrices	May be canonically identified with the group of orthogonal matrices	$\mathfrak{so}(n, \mathbb{R})$	Antisymmetric square matrices; Lie bracket is the commutator	$n(n-1)/2$
$SO(n, \mathbb{R})$	Special orthogonal group of real orthogonal matrices with determinant one	Corresponds to the group $SE(3)$ with a fixed origin, compact	$\mathfrak{so}(n, \mathbb{R})$	Real antisymmetric square matrices	$n(n-1)/2$
$U(n)$	Unitary group of unitary complex $n \times n$ matrices	Isomorphic to $\mathbb{S}^1$ for $n = 1$	$\mathfrak{u}(n)$	Complex square matrices A satisfying $A = -A^*$	n^2

Table A.1. *Examples of Lie groups*

A.4.3. *Representing the flow of a vector field*

After these examples of group transformations, let us briefly return to the question of how to represent and use the flow of a vector field. We need these ideas in order to properly understand E. Cartan's approach to the normalization method presented at the end of this appendix. The description of a flow is related to the infinitesimal generators of transformation groups. Right translation can be used as an example to demonstrate the connection between the two concepts. Right-translation may be identified with the Lie group $G = \mathbb{R}$ via $x \to x + t$. The infinitesimal generator ξ associated with this action is the operator v_ξ of the first-order differential equation $\dot{x} = F(x)$, which satisfies the following relation for every smooth function $F : \mathbb{R} \longmapsto \mathbb{R}$:

$$v_\xi(F(x)) = \frac{d}{dt} F(e^{-t \cdot \xi})|_{t=0}.$$

This gives a clear relation between the flow and the infinitesimal generator (section 1.5.3) that represents this action by an infinitesimal change $v_\xi(F(x))$ at $F(x)$. In other words, $\bar{F}(x) = e^{t\xi} \cdot F(x)$ is the same function but has been reformulated so that the Taylor expansion satisfies:

$$\bar{F}(x) = F(e^{-t \cdot \xi} \cdot x) = F(x) + t.v_\xi(F(x)) + \cdots .$$

In our example, this expansion can be rewritten as: $F(x + t) = e^{(t\partial/\partial x)} F(x)$. Thus, differentiating generates a translation (readers can find further details in [OLV 03]). The flow can be reconstructed directly from the infinitesimal generator by solving a first-order differential equation whose integral curves are the orbits $\mathcal{O}_x = \{e^{-t\xi} \cdot x | t \in \mathbb{R}\}$ that sweep the set of all flows as x ranges over $\mathcal{X}$. Another approach to defining orbits can also be found in the literature: the orbit passing through x may be defined as the set of all images $g \cdot x$ under the action of the group G. For our purposes, flows and orbits are objects that will allow us to qualitatively judge the relevance of a transformation group. Consider for example the isotropic subgroup $G_z = \{g | g \cdot x = x\}$ for $x \in \mathcal{M}$, i.e. the set of all elements that fix x. Transformation groups can have the following properties:

– a group is said to be free if and only if the only element $g \in G$ that fixes any point $x \in \mathcal{M}$ is the identity; $G_x = \{e\}$ for all $x \in \mathcal{M}$;

– a group is said to be locally free if and only if every orbit has the same dimension as G; $G_x \subset G$ is discrete for every $x \in \mathcal{M}$;

– a group is said to be regular if and only if these orbits form a regular stratification.

A.5. Lie algebra associated with a Lie group

As we saw earlier, $L_g : h \longmapsto g \cdot h$ and $R_g : h \longmapsto h \cdot g$ are the charts of left and right multiplication, respectively. The vector field v is said to be right invariant if $DL_g(v) = v$ for all $g \in G$, and left invariant if $DR_g(v) = v$ for all $g \in G$. In other words, for all $h, g \in G$, right multiplication must satisfy $(DR_g)|_h(v(h)) = v(h \cdot g)$, and left multiplication must satisfy $(DL_g)|_h(v(h)) = v(g \cdot h)$. Thus, any right invariant (respectively, left invariant) vector field is uniquely determined by its value at the identity element e, i.e. $v|_g = DR_g(v|_e)$. This allows us to identify the right (respectively, left) Lie algebra with the tangent space of G at the identity element, $TG|_e$. As an example, let us return to the example of the affine group A(1). The following pair of right invariant vector fields give a basis $\partial_a|_e, \partial_b|_e$, of $TA(1)|_e$ for the right-transformation group $R_{(a,b)}(c, d) = (ac, bc + d)$:

$$v_1 = DR_{(a,b)}[\partial_a|_e] = a\partial_a + b\partial_b \quad ; \quad v_2 = DR_{(a,b)}[\partial_b|_e] = \partial_b. \qquad \text{[A.1]}$$

We will use this idea of invariant vector fields in other chapters to construct invariant observers. Whenever a Lie group acts on itself by left or right multiplication, we saw earlier that there exists an invariant vector field, the Lie algebra. For more general group actions, the invariant vector field w is characterized by the condition that $Dg(w|_x) = w|_{g \cdot x}$ for every element $g \in G$ and every x in the domain of g. The infinitesimal invariance criterion for any such vector field can be phrased in terms of the Lie bracket (see [OLV 03] for further details). Compared to the special case of invariant functions or vector fields on Lie groups, it can be difficult to show that a vector field exists for general transformation groups. As an example, consider the group of translations acting on $\mathbb{R}$, $(x \longrightarrow x + b)$. For this group, every constant multiple of the translation of the vector field $\partial/\partial x$ is an invariant vector field. Indeed, arbitrary vector fields can be expressed as $u(x)\partial/\partial x$. By definition, in order for this vector field to be right invariant, we must have that $u(x + t) = u(x)$ for every t. Therefore, every right-invariant vector field must be proportional to the field of constant vectors $\partial/\partial x$. However, for other transformation groups such as "affine" or "projection" groups, we cannot guarantee the existence of invariant vector fields. The only case in which the number of available invariant vector fields is predetermined is the case in which the group acts freely, i.e. the dimension of the space on which the group acts is sufficiently large ($m < r = dim(G)$). In this case, we are guaranteed to have a maximum collection that includes every possible invariant vector field. In cases where the group does not act freely, one possible approach (often difficult to implement in practice) is to find a natural prolongation procedure that increases the dimension of the space on which the group is acting.

A.6. Cartan's method of moving frames

To end this section, we present a key method that is used with Lie transformation groups: normalization. The normalization method lies at the heart of Cartan's approach to moving frames and equivalence problems.

A.6.1. *Example in* $\mathbb{R}^3$

The basic idea is straightforward. Consider a manifold Σ defined as a collection of open subsets $U_\alpha \subset \mathcal{X}$ in a Euclidean space on $\mathbb{R}^3$, together with a two-parameter transformation group G acting on Σ whose orbits are planes (e.g. straightened-out coordinates). Normalization is based on the following idea: provided that the regularity and freedom conditions stated at the end of section A.4.3 are satisfied by the transformation group, we can define a set of coordinates (ξ_1, ξ_2, ξ_3) as shown in Figure A.7; the directions $\xi_f = (\xi_1, \xi_2)$ range over each orbit, whereas the transverse direction $\xi_b = \xi_3$ determines on which orbit we are located. Thus, to determine whether any two points P and S are located on the same orbit, i.e. whether there exists an element $g \in G$ such that, written succinctly, $S = \varphi_{g(P)}(P)$, we can simply check whether these two points have the same ξ_3-coordinate. This can be reformulated as the problem of finding an invariant function for ξ_3. Suppose that the points P, S and R have coordinates (p_1, p_2, p_3), (s_1, s_2, s_3) and (r_1, r_2, r_3) with respect to some given coordinate system. As noted, if P and S belong to the same orbit, then there exists an element $g \in G$ such that:

$$s_1 = (\varphi_{g(P)}(P))_1, s_2 = (\varphi_{g(P)}(P))_2, s_3 = (\varphi_{g(P)}(P))_3.$$

The essence of the problem lies in switching to the local case for the parameters of the group g, i.e. finding the invariant functions of the local transformation group. To find these invariant functions, we need to eliminate the two parameters associated with g. To do this, we introduce a third point R on the same orbit as P to serve as a reference. The ξ_3-coordinate of R

belongs to the group of coordinates parametrized by the transverse section of every orbit such that $K = \{\xi_3^1 = c_1, \cdots, \xi_3^r = c_r\}$, with index $r := dim(G)$. Locally, we can then determine the element $g(P)$ of this group that satisfies $R = \varphi_{g(P)}(P)$. This element is a solution of the following system of equations:

$$r_1 = (\varphi_{g(P)}(P))_1, r_2 = (\varphi_{g(P)}(P))_2.$$

By the implicit function theorem cited in section A.1 (the rank of this system is always two for the two parameters of G), we obtain the mapping $\gamma : \Sigma \to G$, known as the "moving frame":

$$g(P) = \gamma(P, r_1, r_2).$$

Thus, invariant functions can be expressed as the normalization of the transformation group by the moving frame, i.e.:

$$I(P) = (\varphi_{\gamma(P,r_1,r_2)}(P))_3.$$

The invariance property implies that $I(\varphi_g(P)) = I(P)$, and, since R and S are on the same orbit, $R = \varphi_{\gamma(S,r_1,r_2)}(S)$.

A.6.2. *General case*

Consider a group G with a regular action on an n-dimensional manifold Σ with r-dimensional orbits such that $r \leq s$. In local coordinates ξ on $\Sigma \subset \mathbb{R}^n$, write $(\phi_g)_g \in G$ for the transformation associated with the element g. As mentioned in section 1.6, to guarantee the existence of a local solution, we will assume that $\partial_g \phi_g$ has full rank $r := dim(G)$ at the point $(e, z^0) \in G \times \Sigma$. This means that we can decompose ϕ_g into two components (ϕ_g^a, ϕ_g^b) with dimensions r and $n - r$, respectively, such that ϕ_g^a is invertible with respect to g around (e, z^0). We now obtain the normalization equations by setting:

$$\phi_g^a(z) = c,$$

where c is a constant that belongs to the image of ϕ^a. The implicit function theorem guarantees the existence of a local solution (namely the moving frame $\gamma : \Sigma \to G$) such that $g = \gamma(z)$, which is the unique element that maps z to the transverse section. In this way, we obtain a complete set I of $n - r$ functionally independent invariants by substituting $g = \gamma(z)$ into the other components of the transformation,

$$I(z) := \phi^{b}_{\gamma(z)}(z).$$

The method of moving frames thus allows us to find a complete set of local invariants $I(z) \in \mathbb{R}^n$ from the reduction of the group action to its action on the state space $\mathcal{X}$.

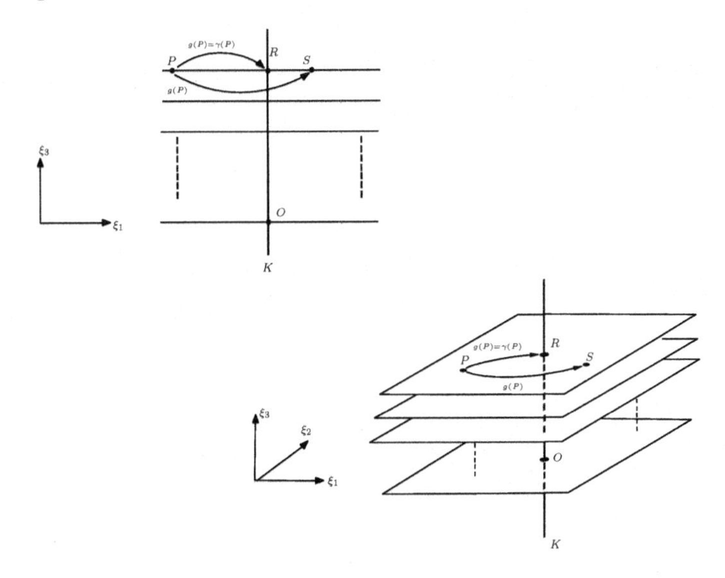

Figure A.7. *Orbits of the transformation group G acting on Σ in straightened-out coordinates. $2D$ and $3D$ cases*

References

[AGH 02] AGHANNAN N., ROUCHON P., "On invariant asymptotic observers", *Proceedings of the 41st IEEE Conference on Decision and Control*, pp. 1479–1484, Las Vegas, USA, December 2002.

[ALD 01] ALDON M.J., "Capteurs et méthodes pour la localisation des robots mobiles", S7852, *Techniques de l'Ingénieur*, 2001.

[ALL 66] ALLAN D., "Statistics of atomic frequency standards", *Proceedings of the IEEE*, vol. 54, pp. 221–230, 1966.

[ALL 87] ALLAN D.W., "Time and frequency (time-domain) characterization, estimation, and prediction of precision clocks and oscillators", *IEEE Transportations on Ultrasonics, Ferroelectrics, and Frequency Control*, vol. 34, pp. 647–654, 1987.

[ALP 72] ALPACH D.L., SORENSON H.W., "Nonlinear Bayesian estimation using Gaussian sum", *IEEE Transactions on Automatic Control*, vol. 17, pp. 439–448, 1972.

[ALP 92] ALPERT B.K., "Wavelets and other bases for fast numerical linear algebra", in CHUI C.K. (ed.), *Wavelets : a tutorial in theory and applications*, Academic Press, San Diego, 1992.

[BAR 96] BAR-ITZHACK I.Y., "REQUEST: a recursive QUEST algorithm for sequential attitude determination", *Journal of Guidance, Control, and Dynamics*, vol. 19, pp. 1034–1038, 1996.

[BAR 13] BARCZYK M., LYNCH A.F., "Invariant observer design for a helicopter UAV aided inertial navigation system", *IEEE Transactions on Control Systems Technology*, vol. 21, pp. 791–806, 2013.

[BAR 14] BARRAU A., BONNABEL S., "The invariant extended Kalman filter as a stable observer", *IEEE Transactions on Automatic Control*, vol. 62, no. 4, pp. 1797–1812, 2014.

[BEL 94] BELL B., "The iterated Kalman smoother as a Gauss-Newton method", *SIAM Journal of Optimization*, vol. 4, pp. 626–636, 1994.

[BER 72] BERGER M., GOSTIAUX B., *Géométrie différentielle: variétiés, courbes et surfaces*, Armand Colin, Paris, 1972.

[BES 83] BESTLE D., ZEITZ M., "Canonical form observer design for nonlinear time-variable systems", *International Journal of Control*, vol. 38, pp. 419–431, 1983.

[BES 04] BESANÇON G., ZHANG Q., HAMMOURI H., "High-gain observer based state and parameter estimation in nonlinear systems", in *Proceedings of NOLCOS, IFAC Symposium on Nonlinear Control Systems*, 2004.

[BIJ 08] BIJKER J., STEYN W., "Kalman filter configurations for a low-cost loosely integrated inertial navigation system on an airship", *Control Engineering Practice*, vol. 16, pp. 1509–1518, 2008.

[BIR 88] BIRK J., ZEITZ M., "Extended Luenberger observer for nonlinear multivariable systems", *International Journal of Control*, vol. 47, pp. 1823–1836, 1988.

[BON 07] BONNABEL S., "Left-invariant extended Kalman filter and attitude estimation", *46th IEEE Conference on Decision and Control*, pp. 1027–1032, 2007.

[BON 08] BONNABEL S., MARTIN P., ROUCHON P., "Symmetry-preserving observers", *IEEE Transactions on Automatic Control*, vol. 53, pp. 2514–2526, 2008.

[BON 09a] BONNABEL S., MARTIN P., ROUCHON P., "Nonlinear symmetry-preserving observers on Lie groups", *IEEE Transactions on Automatic Control*, vol. 54, pp. 1709–1713, 2009.

[BON 09b] BONNABEL S., MARTIN P., SALAÜN E., "Invariant extended Kalman filter: theory and application to a velocity-aided attitude estimation problem", *IEEE Conference on Decision and Control*, pp. 1297–1304, 2009.

[BRE 67] BREAKWELL J.V., Estimation with slight non-linearity, Unpublished communication, 1967.

[BRO 13] BRONZ M., CONDOMINES J.-P., HATTENBERGER G., "Development of an 18 cm micro air vehicle: QUARK", *International Micro Air Vehicle Conference and Flight Competition (IMAV)*, September 2013.

[CON 13] CONDOMINES J.-P., CÉDRIC S., HATTENBERGER G., "Nonlinear state estimation using an invariant unscented Kalman filter", *AIAA Guidance, Navigation and Control (GNC) Conference*, pp. 1–15, 2013.

[CON 14] CONDOMINES J.-P., CÉDRIC S., HATTENBERGER G., "Pi-invariant unscented Kalman filter for sensor fusion", *53rd IEEE Conference on Decision and Control*, pp. 1035–1040, 2014.

[CRA 03] CRASSIDIS J.L., MARKLEY F.L., "Unscented filtering for spacecraft attitude estimation", *Journal of Guidance, Control, and Dynamics*, vol. 26, pp. 536–542, 2003.

[CYG 09] CYGANEK B., SIEBERT J.P., *An Introduction to 3D Computer Vision Techniques and Algorithms*, John Wiley & Sons, 2009.

[DAR 13] DARLING J.E., BALAKRISHNAN S.N., D'SOUZA C., "Sigma point modified state observer for nonlinear unscented estimation", *AIAA Guidance, Navigation and Control Conference*, 2013.

[DEA 13] DE AGOSTINO M., MANZINO A.M., PIRAS M., "Performance comparison of different MEMS-based IMUs", *Position Location and Navigation Symposium (PLANS) IEEE*, 2013.

[DEL 08] DE LAUBIER A., Etalonnage et mise en œuvre d'une centrale inertielle pour la localisation 6D d'un robot mobile, PhD thesis, Conservatoire National des Arts et Métiers, Clermont-Ferrand, 2008.

[DIO 91] DIOP S., FLIESS M., "Nonlinear observability, identifiability, and persistent trajectories", *IEEE Conference on Decision and Control*, pp. 714–710, December 1991.

[DIO 09] DIOP S., SIMEONOV I., "On the biomass specific growth rates estimation for anaerobic digestion using differential algebraic techniques", *Bioautomation*, vol. 13, pp. 47–56, 2009.

[DIO 12] DIOP S., "On a differential algebraic approach of control observation problems", *18th International Conference on Applications of Computer Algebra*, Sofia, Bulgaria, June 2012.

[FAR 98] FARRELL J.A., BARTH M., *The Global Positioning System and Inertial Navigation*, McGraw-Hill, New York, 1998.

[FAR 00] FARUQI F., TURNER K., "Extended Kalman filter synthesis for integrated global positioning/inertial navigation systems", *Applied Mathematics and Computation*, vol. 115, pp. 213–227, 2000.

[FAR 04] FARZA M., SAAD M., ROSSIGNO L., "Observer design for a class of MIMO nonlinear systems", *Automatica*, vol. 40, pp. 135–143, 2004.

[FAR 08] FARRELL J.A., *Aided Navigation GPS with High Rate Sensors*, McGraw-Hill, New York, USA, 2008.

[FIS 22] FISHER R.A., "On the Mathematical Foundations of Theoretical Statistics", *Philosophical Transactions of the Royal Society London A*, vol. 22, nos 594–604, pp. 309–368, 1922.

[GAU 92] GAUTHIER J.P., HAMMOURI H., OTHMAN S., "A simple observer for nonlinear systems applications to bioreactors", *IEEE Transactions on Automatic Control*, vol. 37, pp. 875–880, 1992.

[GOR 93] GORDON N.J., SALMOND D.J., SMITH A.F.M., "Novel approach to nonlinear/non-Gaussian Bayesian state estimation", *IEEE Proceedings on Radar and Signal Processing*, vol. 140, pp. 107–113, 1993.

[GRE 01] GREWAL M.S., WEILL L.R., ANDREWS A.P., *Global Positioning Systems, Inertial Navigation, and Integration*, John Wiley & Sons, 2001.

[HAT 14] HATTENBERGER G., BRONZ M., GORRAZ M., "Using the Paparazzi UAV system for scientific research", *IMAV 2014: International Micro Air Vehicle Conference and Competition*, pp. 12–15, August 2014.

[HIG 75] HIGGINS W.T., "A comparison of complementary and Kalman filtering", *IEEE Transactions on Aerospace and Electonic Systems*, vol. 11, pp. 321–325, 1975.

[JAZ 70] JAZWINSKI A.H., *Stochastic Processes and Filtering Theory*, New York Academic, 1970.

[JUL 95] JULIER S.J., UHLMANN J.K., DURRANT-WHYTE H.F., "A new approach for filtering nonlinear systems", *Proceedings of the American Control Conference*, pp. 1628–1632, 1995.

[JUL 00] JULIER S.J., UHLMANN J.K., DURRANT-WHYTE H.F., "A new method for the nonlinear transformation of means and covariances in filters and estimators", *IEEE Transactions on Automatic Control*, vol. 45, pp. 477–482, 2000.

[JUL 02] JULIER S.J., UHLMANN J.K., "Reduced sigma point filters for the propagation of means and covariances through nonlinear transformations", *Proceedings of the American Control Conference*, pp. 887–891, 2002.

[JUL 03] JULIER S.J., "The spherical simplex unscented transformation", *Proceedings of the American Control Conference*, vol. 3, pp. 2430–2434, 2003.

[JUL 04] JULIER S.J., UHLMANN J.K., "Unscented filtering and nonlinear estimation", *Proceedings of the IEEE*, vol. 92, pp. 401–422, 2004.

[KAL 60] KALMAN R.E., "A new approach to linear filtering and prediction problems", *Transactions of the ASME Journal of Basic Engineering*, vol. 82, no. 1, pp. 35–45, 1960.

[KAL 61] KALMAN R.E., BUCY R.S., "New results in linear filtering and prediction theory", *Transactions of the ASME Journal of Basic Engineering*, vol. 23, no. 1, pp. 95–108, 1961.

[KAZ 98] KAZANTZIS N., KRAVARIS C., "Nonlinear observer design using Lyapunov's auxiliary theorem", *Systems and Control Letters*, vol. 34, pp. 241–247, 1998.

[KEL 87] KELLER H., "Nonlinear observer design by transformation into a generalized observer canonical", *International Journal of Control*, vol. 46, pp. 1915–1930, 1987.

[KOL 22] KOLMOGOROV A., *Grundbegriffe der Wahrscheilichkeitrechung*, Springer, 1922.

[KOU 75] KOU S.R., ELLIOTT D.L., TARN T.J., "Exponential observers for nonlinear dynamic exponential observers for nonlinear dynamic systems", *Information and Control*, vol. 29, 1975.

[KRE 83] KRENER A.J., ISIDORI A., "Linearization by output injection and nonlinear observers", *Systems and Control Letters*, vol. 3, pp. 47–52, 1983.

[LAG 08] LAGEMANN C. J.T., MAHONY R., Observer design for invariant systems with homogeneous observations, Preprint, arXiv, 2008.

[LAU 01] LAUMOND J.P., *La robotique mobile*, Hermès, Paris, 2001.

[LAV 04] LAVIOLA J.J., "A comparison of unscented and extended Kalman filtering for estimating quaternion motion", *Proceedings of the 2004 American Control Conference*, pp. 2190–2195, 2004.

[LAW 98] LAWRENCE A., *Modern Inertial Technology: Navigation, Guidance, and Control*, Springer-Verlag, 1998.

[LEF 82] LEFFERTS E.J., MARKLEY F.L., SHUSTER M.D., "Kalman filtering for spacecraft attitude estimation", *Journal of Guidance, Control and Dynamics*, vol. 5, pp. 417–429, 1982.

[LEF 02] LEFEBVRE T., BRUYNINCKX H., DE SCHUTTER J., "Comment on 'A new method for the nonlinear transformation of means and covariances in filters and estimators'", *IEEE Transactions on Automatic Control*, vol. 47, no. 8, pp. 1406–1409, 2002.

[LEV 86] LEVINE J., MARINO R., "Nonlinear system immersion, observers and finite dimensional filters", *Systems and Control Letters*, vol. 7, pp. 133–142, 1986.

[LIM 12] LIM H., PARK J., "Open-source projects on unmanned aerial vehicles", *IEEE Robotics and Automation*, vol. 19, pp. 1070–9932, 2012.

[LUE 71] LUENBERGER D.G., "An introduction to observers", *IEEE Transactions on Automatic Control*, vol. 16, pp. 596–602, 1971.

[MA 04] MA Y., SOATTO S., KOSECKA J. *et al.*, *An Invitation to 3-D Vision: From Images to Geometric Models*, Springer, 2004.

[MAH 08] MAHONY R., HAMEL T., PFLIMLIN J.-M., "Nonlinear complementary filters on the special orthogonal group", *IEEE Transaction Automatic Control*, vol. 53, pp. 1203–1218, 2008.

[MAI 05] MAITHRIPALA D., DAYAWANSA W.P., BERG J.M., "Intrinsic observer-based stabilization for simple mechanical systems on Lie groups", *SIAM Journal on Control and Optimization*, vol. 44, pp. 1691–1711, 2005.

[MAN 12] MANECY A., VIOLLET S., MARCHNAD N., "Bio-inspired hovering control for an aerial robot equipped with a decoupled eye and a rate gyro", *Proceedings of the IEEE/RJS International Conference on Intelligent Robots and Systems*, pp. 1110–1117, 2012.

[MAR 93] MARKLEY F.L., BERMAN N., SHAKED U., "H_∞-type filter for spacecraft attitude flitering", *AAS/GSFC International Symposium on Space Flight Dynamics*, vol. 84, no. 1, 1993.

[MAR 07] MARTIN P., SALAÜN E., "Invariant observers for attitude and heading estimation from low-cost inertial and magnetic sensors", *IEEE Conference on Decision and Control*, pp. 1039–1045, 2007.

[MAR 08] MARTIN P., SALAÜN E., "An invariant observer for earth-velocity-aided attitude heading reference systems", *17th World Congress of the International Federation of Automatic Control*, vol. 41, no. 2, pp. 9857–9864, 2008.

[MAR 10] MARTIN P., SALAÜN E., "Design and implementation of a low-cost observer-based attitude and heading reference system", *Control Engineering Practice*, pp. 712–722, 2010.

[MEH 70] MEHRA R.K., "On the identification of variances and adaptive Kalman filtering", *IEEE Transactions on Automatic Control*, vol. 15, pp. 175–184, 1970.

[MOO 75] MOOSE R.L., "An adaptive state estimation solution to the maneuvering target problem", *IEEE Transactions on Automatic Control*, vol. 20, pp. 359–362, 1975.

[MOO 87] MOOSE R.L., SISTANIZADEH M.K., SKAGFJORD G., "Adaptive state estimation for a system with unknown input and measurement bias", *IEEE Journal of Oceanic Engineering*, vol. 12, 1987.

[MUR 78] MURRELL J.W., "Precision attitude determination for multimission spacecraft", *AIAA Guidance and Control Conference*, 1978.

[NAG 91] NAGPAL K.M., KHARGONEKAR P.P., "Filtering and smoothing in an H_∞ setting", *IEEE Transactions on Automatic Control*, vol. 36, pp. 152–166, 1991.

[NWO 07] NWOGUGU M., "Decision-making, risk and corporate governance – a critique of methodological issues in bankruptcy/recovery prediction models", *Mathematics and Computation*, pp. 178–196, 2007.

[OLV 03] OLVER P.J., *Classical Invariant Theory*, Cambridge University Press, 2003.

[OLV 08] OLVER P.J., *Equivalence, Invariance and Symmetry*, Cambridge University Press, 2008.

[PIC 91] PICARD J., "Efficiency of the extended Kalman filter for nonlinear systems with small noise", *Applied Mathematics and Computation*, vol. 51, pp. 359–410, 1991.

[PSI 00] PSIAKI M.L., "Attitude-determination filtering via extended quaternion estimation", *Journal of Guidance, Control, and Dynamics*, vol. 23, pp. 206–214, 2000.

[RAÏ 10] RAÏSSI T., VIDEAU G., ZOLGHADRI A., "Interval observer design for consistency checks of nonlinear continuous-time systems", *Automatica*, vol. 46, no. 3, pp. 518–527, 2010.

[RAN 07] RANDAL R.W., "Estimation for micro air vehicles", *Studies in Computational Intelligence*, vol. 70, pp. 173–199, 2007.

[RES 85] RESPONDEK W., KRENER A.J., "Nonlinear observer with linearizable error dynamics", *SIAM Journal on Control and Optimization*, vol. 23, pp. 197–216, 1985.

[RUF 04] RUFFIER F., FRANCESCHINI N., "Visually guided micro-aerial vehicle: automatic take off, terrain following, landing and wind reaction", *Proceedings of the IEEE International Conference on Robotics and Automation*, vol. 3, pp. 2339–2346, 2004.

[RUF 09] RUFFIER F., SERRES J., PORTELLI G. *et al.*, "Boucles visuo-motrices biomimétiques pour le pilotage automatique de micro-aéronefs", *7ème Journées Nationales de la Recherche en Robotique*, Proceedings, 2009.

[SAR 07] SARKKA S., "On unscented Kalman filtering for state estimation of continous-time nonlinear systems", *IEEE Transactions on Automatic Control*, vol. 52, pp. 1631–1641, 2007.

[SED 93] SEDLAK J., CHU D., "Kalman filter estimation of attitude and gyro bias with the QUEST observation model", *AAS/GSFC International Symposium on Space Flight Dynamics*, 1993.

[SHU 81] SHUSTER M.D., OH S.D., "Three-axis attitude determination from vector observations", *Journal of Guidance and Control*, vol. 4, 1981.

[SHU 89] SHUSTER M.D., "A simple Kalman filter and smoother for spacecraft attitude", *Journal of the Astronautical Sciences*, vol. 38, pp. 377–393, 1989.

[SMI 95] SMITH R.H., "An H_∞-type filter for GPS-based attitude estimation", *AAS/AIAA Space Flight Mechanics Conference*, Albuquerque, USA, Feb 1995.

[STE 01] STEIMANN F., ADLASSNIG K.P., "Fuzzy medical diagnosis", available at: http://citeseer.nj.nec.com/160037.html, 2001.

[SWE 59] SWERLING P., "First order error propagation in a stagewise smoothing problem for satellite observations", *Journal of Astronautical Science*, vol. 6, no. 3, pp. 46–52, 1959.

[THA 73] THAU F.E., "Observing the state of nonlinear dynamic systems", *International Journal of Control*, vol. 17, pp. 471–479, 1973.

[TIT 04] TITTERTON D.H., WESTON J.L., *Strapdown Inertial Navigation Technology*, The Institution of Electrical Engineers, 2004.

[TRU 98] TRUCCO E., VERRI A., *Introductory Technique for 3-D Computer Vision*, Prentice Hall, 1998.

[VAN 04] VAN DER MERWE R., Sigma-Point Kalman Filters for probabilistic inference in dynamic state-space models, PhD thesis, OGI School of Science and Engineering, Oregon Health and Science University, April 2004.

[VAS 08] VASCONCELOS J., SILVESTRE C., OLIVEIRA P., "A nonlinear observer for rigid body attitude estimation using vector observations", *17th World Congress of the International Federation of Automatic Control*, pp. 8599–8604, 2008.

[VIT 67] VITERBI A., "Error bounds for convolutional codes and an asymptotically optimum decoding algorithm", *IEEE Transations on Information Theory*, pp. 260–269, 1967.

[WAN 00] WAN E., VAN DER MERWE R., "The unscented Kalman filter for nonlinear estimation", *Proceedings of Symposium 2000 on Adaptive Systems for Signal Processing, Communication and Control (AS-SPCC)*, pp. 153–158, 2000.

[WAN 01] WAN E., VAN DER MERWE R., *The Unscented Kalman Filter, Kalman Filtering and Neural Networks*, John Wiley & Sons, 2001.

[WEN 03] WENGER L., GEBRE-EGZIABHER D., "Systems concepts and performances analysis of multi-sensor navigation system for UAV applications", *2nd AIAA Unmanned Unlimited Systems Conference*, 2003.

[XU 08] XU J., WANG S., DIMIROVSKI G. *et al.*, "Stochastic stability of the continuous-time unscented Kalman filter", *IEEE Conference on Decision and Control*, pp. 5110–5115, 2008.

[ZAD 65] ZADEH L.A., "Fuzzy sets", *Information and Control*, vol. 8, no. 3, pp. 338–353, 1965.

[ZEI 88] ZEITZ M., "Comments on comparative study of nonlinear state observation techniques", *International Journal of Control*, vol. 46, pp. 1823–1836, 1988.

Index

Printed in the United States
By Bookmasters